PLANT WORDS

Ania!
To the most dedicated volunteer! It's always a pleasure.
Joe x

Published in 2022 by Welbeck
An imprint of Welbeck Non-Fiction Limited,
part of Welbeck Publishing Group.

Based in London and Sydney

www.welbeckpublishing.com

A CIP catalogue record for this book is available from the British Library.

ISBN 978-1-80279-008-5

Printed in the UK

10 9 8 7 6 5 4 3 2 1

Royal Botanic Gardens Kew

PLANT WORDS

A book of 250 curious words for plant lovers

JOE RICHOMME & EMMA WAYLAND

Contents

Introduction

It is possible to describe or explain just about anything using generic words that most adults would understand. But to really delve into a subject, it becomes necessary to start using specialist terms to make it easier and more efficient to talk about something. Why take a sentence to say what you mean, when one word will do?

The use of specialist terms presents a problem, however. To take part in the conversation, you must know what these words really mean, and it is easy for those who use such terms on a regular basis to simply throw them into conversation without explanation, as if everyone will automatically understand them. Plants are a subject particularly prone to this issue. The botanical world is one where science, industry, hobby, history and culture all intersect. Words have crossed the boundaries between these areas, accruing meaning and/or being misapplied as they go. The importance of places, people and concepts have come to be taken as a given.

This book is intended to untangle some of those assumptions, to try and explain some specialist words concerning plants in relatively simple terms, to convey what they are really about, and to highlight how what they relate to is important. It is an introduction to the vocabulary used by botanists, horticulturists, and many others who interact with plants on a regular basis. It is intended to show you the way deeper into the world of plants; words are the way into any subject.

We have grouped terms together into key topic areas. In **Botany** we discuss those words which describe parts of plants: What is a bulb and how does it differ from a corm?

Where does the word "pistil" come from? In **Growing** we examine words used in our direct interactions and interventions with plants: How do we go about coppicing something? What do we mean by "mulch"? In **Plant Types** we look at different kinds of plants, whether distinguished by their life cycle (annual, perennial), by their natural habitat type (alpine, aquatic) or by their properties (medicinal, hallucinogenic). In **History** we tell some important and interesting stories about plants in relation to the progress of human society. It is worth noting that we focus here on the history of plants, not gardens. **Documentation** examines how we record information about plants, and how that information is used. By **Environment & Ecology** we mean the world within which plants exist – how do they interact with their surroundings? What words do we need to know to consider plants within their wider communities? This is distinct from **Biomes & Habitats**, which looks at the kind of places where plants are found. Finally, in **Science** we look at some of the techniques which scientists use to explore the world of plants, and what scientific examination has taught us about these fascinating organisms.

Unfortunately, this book cannot be exhaustive. We have had to be selective in our choice of terms. We have picked words that might commonly arise in any discussion of plants, and words that perhaps should be part of the conversation. Words which are often confused in their meaning, and words which everyone thinks they know the meaning of but are perhaps worth looking at again. Even words which are perhaps a little obscure but which we think are extremely interesting. In short, this is a personal selection, and a wide-ranging one at that. It is a curiosity cabinet of botanical anatomy and evolutionary innovation; a chronicle of plants in human history; an atlas of our planet's incredible diversity and a warning about its fragility; a tribute to the complexity and importance of modern scientific thinking.

In talking about words, it can be useful to think about where they come from. Understanding the origin of a word can help recall and understanding. If you know that the word "bonsai" comes from the Japanese words for "basin" and "planting", then you quickly get a sense of what that word really means. But sometimes a word has simply existed in almost the same form for millennia. While it is perhaps striking that grass has been called something very similar since the earliest Proto-Indo-European languages, it doesn't provide much insight into the thing itself. In other cases, the root of the word or phrase is so patently obvious that it seems frankly patronizing to labour the point. So, etymology is discussed, but only where it makes meaning clearer, or is particularly interesting in its own right.

All of this must be qualified; this is not a reference book. Concepts, proofs and ideas have had to be simplified to stop the text becoming too academic and unwieldy. In particular, several concepts discussed are applicable to a much wider field than just plants but, in the interests of brevity, we have largely restricted these entries to their relevance to the plant kingdom. A great many of the terms discussed have entire books dedicated to them. But we have had to handpick those areas which we think are most interesting, and most likely to make you want to explore further.

Every single one of these entries merely scrapes the surface of a fascinating topic. Each should be seen as a gateway – there is much more to learn than what is contained within these pages, and equally there are many more terms and concepts which branch off from the ones we have chosen, and which are equally fascinating. While we hope that our selection will equip you with some key words to help you talk about plants in more detail, it is intended above all as the start of a personal journey, inspiring you to search out more information about these fascinating organisms upon which we rely so heavily. For through the breadth and implications of these words does it become apparent just how dependent on plants our way of life is.

Note: within definitions, some words are marked in **bold**: these words have their own definitions elsewhere in the book.

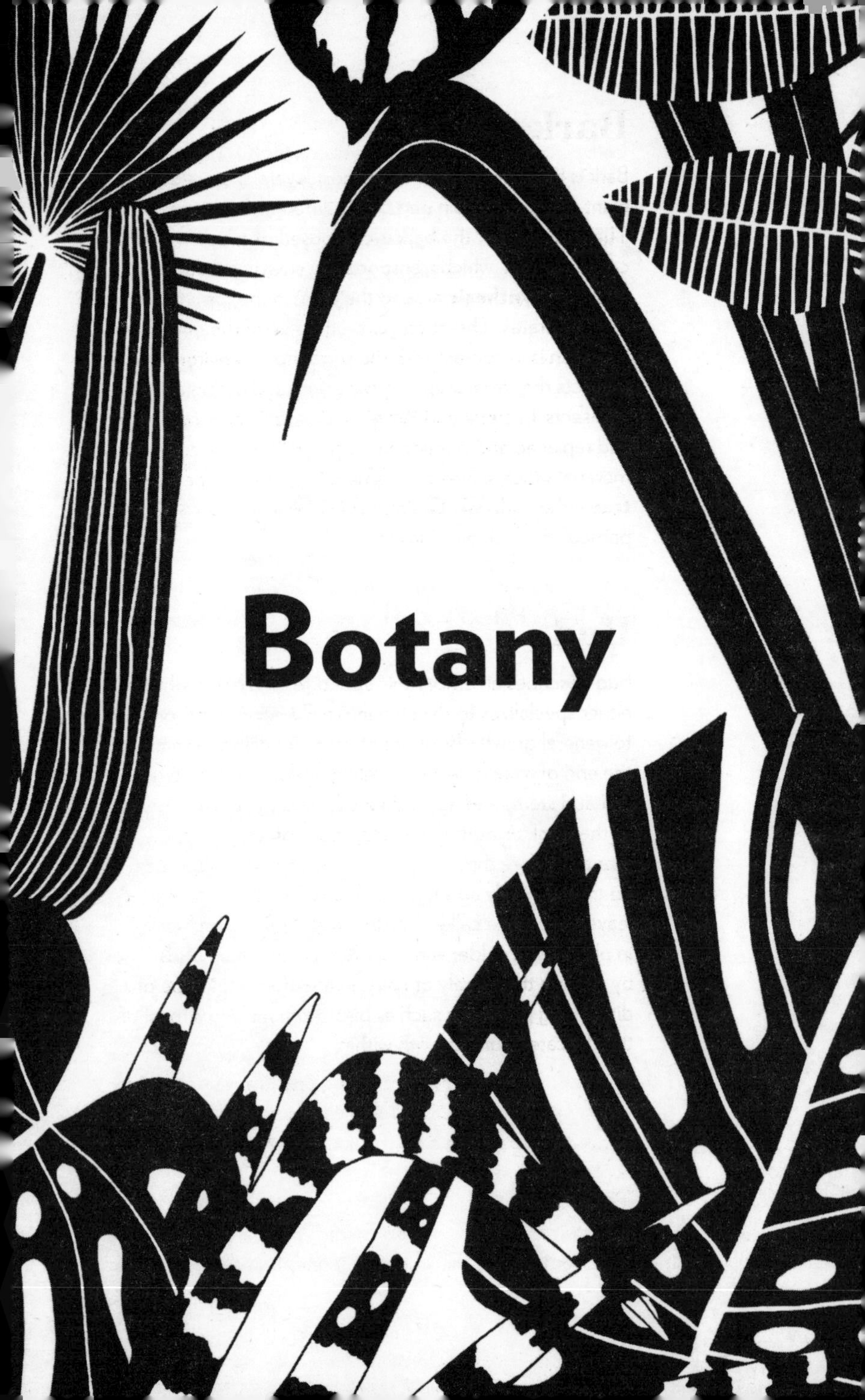

Botany

Bark

Bark is the name for the outermost layers of any woody plant. It has two main parts, with different functions. The inner part of the bark is composed of specialized tissue called phloem, which transports the energy produced by **photosynthesis** around the plant in the form of carbohydrates. The outer part – the skin of the plant, the bit which is in contact with the surrounding environment – protects the stem, stopping water loss and acting as a barrier to insects, bacteria and **fungi**. This layer is continually shed and replaced and is made up of tissue known as cork. The material which shares this name is the bark of a particular **tree** – the cork oak, *Quercus suber*, which produces a particularly thick outer layer.

Bud

Bud describes an embryonic structure on a plant, which either specializes to develop into a **flower** or **leaf**, or leads to general growth. Buds are often divided into terminal (at the end of a set of leaves), axillary (in the angle between leaf and stem) and adventitious (shooting up from the base of the trunk or from the **roots**). Buds often form many months before they will begin to open, and can be seen on trees by early autumn the year before they will produce leaves: these are called "resting buds". Plants, especially in places with colder winters, often protect their buds by making them scaly or hairy, in the often vain hope of dissuading predators such as birds from making a meal of the delicate young growth within.

Bulb

A bulb is an underground plant organ produced by certain plants in which **nutrients** and water are stored. The word comes from the Greek for "onion", which is the perfect example of a bulb's defining characteristic: its modified **leaves** that store water and nutrients, and which are the layers you see when you cut into an onion.

Some bulbs, like tulips, last for only one year – they use up their reserves and their primary growing point producing a **flower**. Any flowers that do appear in subsequent years will actually be from new bulbs formed vegetatively underground and these are always weaker. Bulbs like daffodils will flower well year after year: they produce flowers from secondary buds, which means that the main **bud** is free to produce leaves and top up the plant's energy stores, allowing it to thrive for multiple seasons.

Carpel

Indicating the female reproductive organs of a flowering plant, the term carpel comes from the French *carpelle*, from the Greek *karpos*, meaning "fruit" (which also made its way into Latin as *carpere*, to "cut", "divide" or "pluck", hence *carpe diem*, "seize the day"). The carpel comprises the stigma, the central pillar of a flower which later matures into the **seed** capsule, the ovary and the style, although in different plant families it may appear quite differently. If a flower is cut, the ovules are often visible within the ovary waiting to be fertilized by **pollen**, which will be caught initially on the sticky surface of the stigma, then typically travels down the style to the ovary. Once ripened, the carpel becomes the **fruit**, and a fruit's edible outer layer is hence called a pericarp.

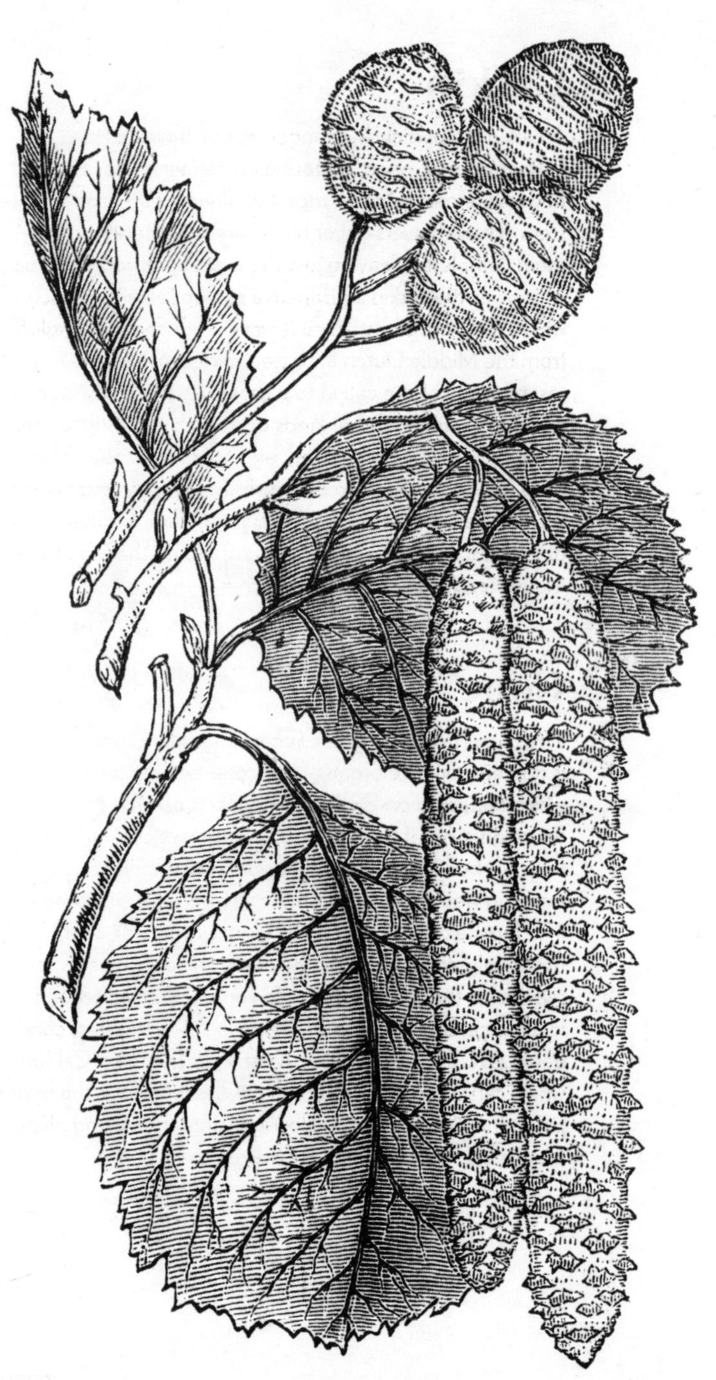

Catkin

A catkin is a particular arrangement of **flowers** (see **Inflorescence**, page 25) found on certain **trees**. Tiny flowers are packed close together along a slim central flower stem. These flowers either have very small petals or lack them completely, leaving just the delicate reproductive parts. This gives the catkin a distinctive soft appearance, which is where the word stems from: it was introduced into English from the Middle Dutch for kitten, *katteken*, because of the resemblance of the catkin to a cat's fur, particularly on certain **species** of willow. Other kinds of tree which produce catkins include alders, birches, hazels, and sweet chestnuts. The male flowers on these trees are always borne in catkins, but the female flowers may be solitary or held in another form of inflorescence. The great majority rely on the wind to blow the **pollen** from one tree to another.

Cone

The cone (from the Greek *konos*, meaning "pinecone") is the reproductive organ of the cone-bearing plants which appeared on Earth many millions of years before the **evolution** of the flower. The **conifers** evolved in the Late Carboniferous age, alongside a set of pollinators, evolving woody cones to protect their ovules and **seeds**. The female cone is heavier and more structural, containing the developing seeds, and may take one or two years to mature on the branch; the male cone is often softer and less noticeable, distributing the pollen. The heaviest cone belongs to the Coulter pine, a native of Southern California, which can weigh up to 5kg; the smallest conifer is the pygmy pine, from New Zealand, a low-growing **alpine** bog plant which very rarely grows taller than 30cm.

Corm

A corm is an underground swollen stem produced by certain plants, which acts as a store of water and **nutrients**. Like **tubers** and **bulbs**, a corm allows plants to survive unfavourable conditions. A corm can look very much like a bulb but does not have layers when cut open. This is because the bulk of a corm (where the nutrients and water are stored) is modified stem tissue, rather than **leaves** like in a bulb. Well-known examples of corm-producing plants include crocus, gladioli and cyclamen, but by far the most impressive is the titan arum – or corpse flower – which produces gigantic corms, the heaviest ever recorded weighing 153.9kg (339lb 4oz).

Crown

The crown of a plant is where the stem joins the **root**. It marks the place where the plant's cells will change from being suited to life above ground to being specialized root cells, adapted to absorb nutrients and moisture from the **soil**. Problems can be caused with planting if the crown is placed too low in the soil, because this may lead to crown rot. However, there are a few plants where gardeners are actually advised to plant them with the crown below the surface, such as potatoes, asparagus and peonies, plus all garden bulbs. When plants are **grafted**, with one plant's above-ground structure mounted onto another's roots, the crown marks where the graft has taken place.

Cuticle

The cuticle is a relatively thin waxy protective layer, covering the leaves and young shoots of land plants, guarding them from environmental or pest damage. The cuticle's primary function is to keep water within the plant at a stable level, but it also helps to prevent pathogenic micro-organisms such as viruses, bacteria and **fungi** from entering the plant. The **evolution** of waxy cuticles was one of the major features that allowed plants to make the transition from life in the water to life on land, some 450 million years ago. Plants that live in water have a special challenge to control water content, and this has led to "the lotus effect", as seen on lotus leaves, where the leaves possess ultrahydrophobicity, beading with water droplets when they come into contact with liquids.

Embryophyte

Embryophyte is the botanical term for the land plants, a group that includes the worts, the **mosses**, the **ferns**, the gymnosperms and the flowering plants (a few members of the group have returned to the water to live). Embryophytes are thought to have evolved from the green **algae** during the beginnings of life on Earth, and they are complex multicellular organisms with a strong cell wall made of cellulose, and usually a cell vacuole that acts as a reservoir to maintain pressure, pH and levels of waste products in the rest of the cell.

Epiphytic

Epiphytic comes from the Greek, from *epi* meaning "in addition", and *phyton* meaning "plant". It means any plant that grows upon another, not parasitically but using the host plant's physical structure for support, and making use of rainwater, dew or mist, and **nutrients**, in the form of plant rubbish, which come to it by chance. They are sometimes called air plants because they do not need to root in **soil**, although many of them have evolved with water reservoirs at their centre, which are used by small **forest** animals such as frogs. Probably the most spectacular epiphytes are the many **species** of Orchidaceae, the orchid **family**, but **lichens** and **mosses** are also epiphytic. They have a strong effect on their local environment via their water reserves, making it cooler and more moist.

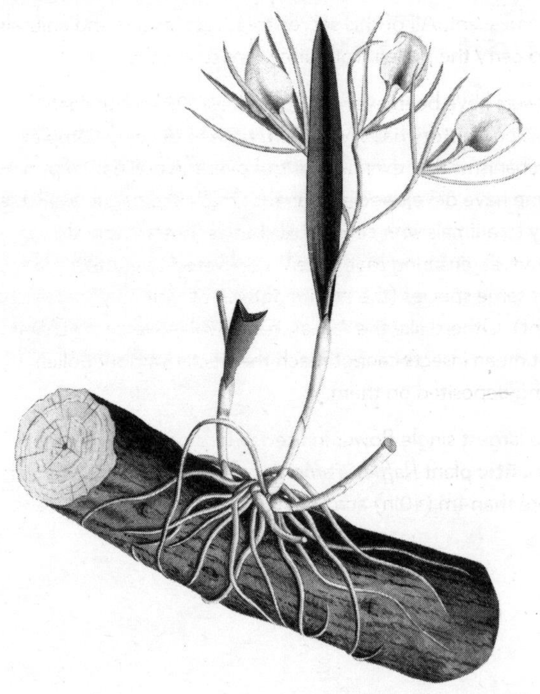

Flower

Flowers are structures found on certain plants. They bear reproductive organs and play a role in **pollination**, **seed** and fruit development. Not all plants produce flowers. Those that do subsequently produce fruits, and so are called angiosperms (see **Fruit**, page 22).

While the most important parts of a flower are the reproductive organs – the **stamen** (male), and the **pistil** (female) – the flower as a whole plays a vital role in reproduction. Plants are not able to move and so, in order to breed with each other, a mechanism for the transfer of genetic material has evolved: **pollen**. Flowers have evolved to make that transfer of pollen between plants more likely, and nowhere more clearly than when this transfer is facilitated by other organisms. They often sport brightly coloured petals in fantastic shapes; they can produce **nectar**, oil and scent. All of this serves to attract insects and animals, who carry the pollen from one plant to another.

Flowers have been evolving for at least 125 million years. This long history has given them time to develop complex mechanisms that exercise control over the pollination process. Some have developed shapes that make the nectar available only to animals with certain bill shapes (hummingbirds, for instance), ensuring that pollen is delivered to another plant of that same species (the hummingbird will visit only that kind of plant). Others, like the salvias, have evolved lever mechanisms that mean insects cannot reach the nectar without pollen being deposited on them.

The largest single flower in the world is produced by the **parasitic** plant *Rafflesia arnoldii*, which can produce flowers more than 1m (40in) across.

Fruit

A fruit is an organ produced by certain plants. Developing from the ovary of the **flower** after fertilization, fruits mature along with the **seeds** and act as their container. Those plants which produce fruits are called angiosperms, from the Greek words for vessel (*angeion*) and seed (*sperma*).

Fruits have primarily evolved as a means of dispersal – the way in which a plant's seeds are spread further afield than its immediate surroundings. The word itself comes from the Latin *fructus*, which also meant "enjoyment", and a common mechanism is that the fruit exists as an enticement; animals seek out the tasty, nutritious fruit, and ingest the seeds within at the same time. The seeds are then carried away in the animal's gut before being deposited elsewhere when the animal defecates.

The term *fruit* is often used in general conversation to mean just those sweet fruits eaten by humans. But the botanical definition takes in a much wider range of forms and tastes. Tomatoes, chillies and pumpkins are all botanical fruits that are eaten savoury, for example. Some nuts like chestnuts and hazelnuts are, botanically speaking, fruits. And ingestion is only one mechanism of dispersal. There are also those fruits which haven't evolved to be eaten. The seed capsule of a poppy, the winged samara of a sycamore, the fluffy parachute of a dandelion; these are all botanical fruits equally adapted to spread a plant's seed, just not by being ingested.

THIEBAULT

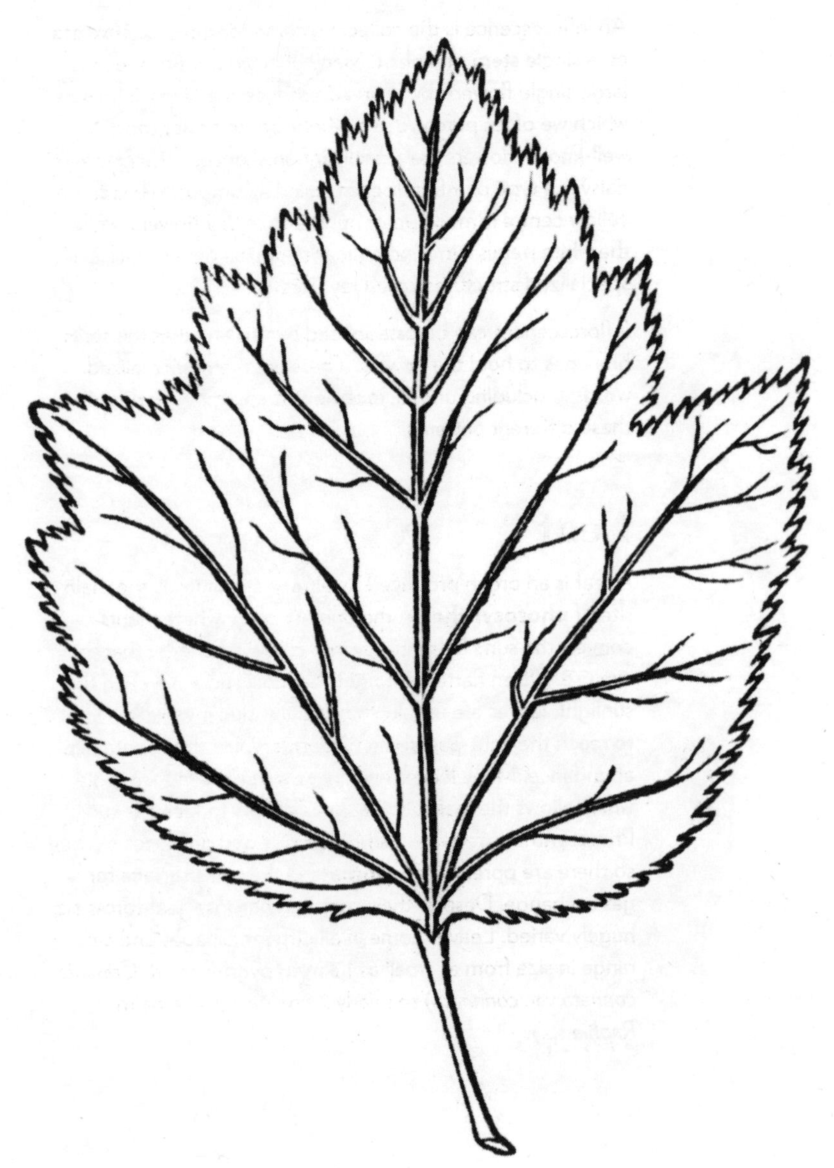

Inflorescence

An inflorescence is the collective name for multiple **flowers** on a single stem of a plant. Many plants do not produce large single flowers, but instead produce many small flowers, which we often perceive as a whole. Some of our most well-known flowers are actually inflorescences. The common daisy is a type of inflorescence called a composite head: the yellow centre is made up of hundreds of tiny flowers, while the white petals surrounding it are not true petals at all, but specialized structures called ray florets.

Inflorescences can be categorized by the way that the stem branches to hold the flowers. There are many specialized words – including umbel, raceme and corymb – to describe these different patterns.

Leaf

A leaf is an organ produced by almost all plants as the main site of **photosynthesis**; the primary place where plants convert the sun's light energy into chemical energy that they can use. Often flattened to get the most surface area in the sunlight, leaves are usually thin and translucent to allow light to reach the light-absorbing pigments which they contain in abundance. Many leaves even have a stem called a petiole, which allows the leaf to orient itself always to face the sun. Photosynthesis requires carbon dioxide and produces oxygen so there are pores called **stomata** in the leaf's surface for gas exchange. Despite their common purpose, leaf forms are hugely varied. Leaves come in all different shapes and can range in size from as small as 1.3 mm (pygmy weed, *Crassula connata var. connnata*) to nearly 20 m long (raffia palm, *Raphia* sp.).

Nectar

Nectar is a sweet liquid produced by plants. It is generally found within **flowers** where it acts as a lure to encourage **pollination**; insects and animals come to the flower to drink the nectar, and so come into contact with the **stamens** and **pistils**, transferring **pollen** between plants as they visit multiple flowers. Nectar is the primary ingredient used by bees to make honey. It is the source of honey's sweetness because it primarily consists of sugars. But it will also contain trace amounts of various other substances including essential oils and pollen, which is why honey can taste different depending on the flowers from which the nectar is collected.

In Ancient Greek myth, nectar was the name for the drink of the gods. The word later came to mean any sweet drink or delicacy, so it was only a short leap for it to be applied here.

Node

In botanical terms, a node is a place on a plant's stem from which **leaves**, **buds**, flowers or **cones** can grow. It is a growth zone, with cells specially adapted to be ready to spring into growth. Because **cactus** spikes are heavily adapted leaves, their points of attachment are also nodes. A **rhizome** is an underground modified plant stem, as opposed to a true **root**, so it also roots and shoots from its nodes. The space in between the nodes is called the internode, and if this area is becoming untypically long, that may mean the plant is not getting enough light. Any **pruning** should consider where a plant's nodes are, as new growth always comes from there.

Pistil

Pistils are the female reproductive parts of a plant. They are usually located in the centre of a **flower**. A pistil is made up of one or more **carpels**, which is the collective term for an ovary, style and stigma. The ovary, as in a human, contains the female reproductive cell – an egg – and this is where the **seed** will develop. The style and stigma facilitate delivery of the male reproductive cell to the egg for fertilization. This process is known as pollination.

The word has its origins in the Latin word *pistillum,* meaning "pestle" (as in pestle and mortar). It was probably appropriated for the botanic meaning because of the pistil's resemblance to the clublike end of Roman pestles.

Pneumatophore

Roots require oxygen, despite usually being underground structures, and this is in short supply in wet or marshy ground. For this reason, some water-dwelling or hydrophytic plants have evolved pneumatophores, vertical roots that grow upwards into the air. From the Greek *pneumato*, meaning "relating to the air", and *-phore*, meaning "carrier", these woody structures allow for gaseous exchange even when the **soil** is completely waterlogged and so low in oxygen. The roots have tissue containing small pores, called lenticels, to allow air into the spongy root cells, and are particularly typical of plants in mangrove and other **tropical** coastal **swamps**. The roots are adapted to grow upwards, rather than downwards like normal roots, by responding inversely to gravity, a process known as negative geotropism.

Pollen

Pollen is a substance that is produced by the male reproductive parts of a plant, and which carries the means to produce sperm cells for fertilization. The word comes from the Latin for "fine flour" and pollen is only visible as a powdery substance, small enough to be blown by the wind or transported easily by insects or animals. This is an adaptation that allows plants to breed with each other despite being non-mobile.

Pollen grains can say a lot about a plant: many plants can be identified from the characteristics of their pollen alone, and similarities between plant pollens can tell us how they might be related to each other. To see this sort of detail, however, the pollen must be magnified many thousands of times with specialist microscopes, giving an idea of just how small a pollen grain is.

Resin

Resin comes from the Greek *rhetine*, for "resin of the pine", which also gives its name to the popular drink retsina. Plant resins are sticky protective substances issued in response to injury and also to protect vulnerable structures such as the **cone**. They are also sometimes produced by plants as **pollination** rewards, and bees may collect them to reinforce their nests. Resins often have a highly recognizable scent, because of their high levels of compounds called terpenes, which have a distinctive quality to the human nose. Members of the plant resin group include asafoetida, balsam, frankincense, myrrh and gum, and chemicals refined from resins include turpentine, shellac and rosin. Their role in the production of incense and perfume, both during life and for embalming at the moment of death, means they have played a role in almost all human history.

Rhizome

A rhizome is a horizontally growing underground stem. Although the word comes from the Ancient Greek for "**root**", a rhizome is structurally different from a root because it has **nodes** – or **buds** – at intervals along its length. From these points it produces both the roots and shoots of new plants, and so acts as a way for the plant to reproduce. Because a rhizome can grow indefinitely, many rhizomatous plants spread very readily.

Many plants have also developed adapted rhizomes, which act as stores for **nutrients** and water, allowing the plant to survive unfavourable conditions. The swollen stems of edible ginger are a perfect example of this kind of rhizome.

Root

A root system is generally the part of a plant that absorbs water and nutrients, and which also serves to anchor a plant in place. While the word stems from the Old Norse *rot*, which referred to the underground parts of a plant, roots do not have to be underground. Certain orchids, for instance, have aerial roots which absorb moisture and minerals directly from the air. Other plants have developed adapted roots for other purposes such as support or aeration.

The tip of a root is protected by a cap of cells which protect the growing point and thus enable the root to push its way through the **soil**. It is thought that the deepest **tree** roots in the world – those of a wild fig tree in South Africa – have penetrated to 120m (395ft) below the surface.

Sap

Sap is the fluid that transports water, sugars, **hormones**, minerals and **nutrients** around the plant in its **vascular** system. It also thus plays a role in sending information around the system about what is happening outside the plant, from the plant's local competitors to daylight length to possible predators. The **evolution** of sophisticated membrane systems has allowed the plant to sort which particular components of the sap it wants to send to which places, and as with many other parts of a plant's existence, recent scientific research has focused on how sap might act as a symbiotic reward for co-evolved animals such as cicadas and aphids, which may benefit the plant by deterring predators. Animals that eat sap often produce honeydew, a sticky, sugary substance which is itself harvested, by ants, wasps and even some **species** of birds.

Seed

A seed is a structure produced by certain plants during sexual reproduction. After fertilization, a new plant begins to develop. In seed-bearing plants, this embryo is encased along with a food store within a skin called a seed coat. Once this seed develops a certain amount, it becomes dormant. This pause allows the new plant to be dispersed, reducing competition between generations, and additionally spreading the plant's DNA. It also ensures that the new plant has the best chance of survival; seeds have evolved so that development continues only when conditions are favourable (see **Germination**, page 48). And a seed can remain viable for a very long time. In 2005, scientists successfully germinated a date seed found during excavations of the fortress of Masada in Southern Israel, a seed that they determined was around 2,000 years old.

Spore

A spore (from the Greek *spork*, to "seed" or "sow") is a piece of reproductive material produced by some plants, and also by **fungi**. The **mosses**, liverworts, club mosses, horsetails and **ferns** all make spores. Although spores can just about be seen by those with good eyesight, they are microscopic in structure, and display an incredible variety of texture and shape when viewed under magnification. Reproduction by this method evolved before plants had developed strong relationships with animal pollinators, and spores are for the most part dispersed by the wind. Because they are very simple structures in comparison to **seeds**, they can remain viable even after many dry months, or even years. Strong spore walls create clear fossils, whose presence in the record is seen as a sign of the arrival of the first land plants, about half a billion years ago. In Canning Basin, Western Australia, 480-million-year-old fossilized spores have been found, predating by at least 35 million years any fossil of a whole plant.

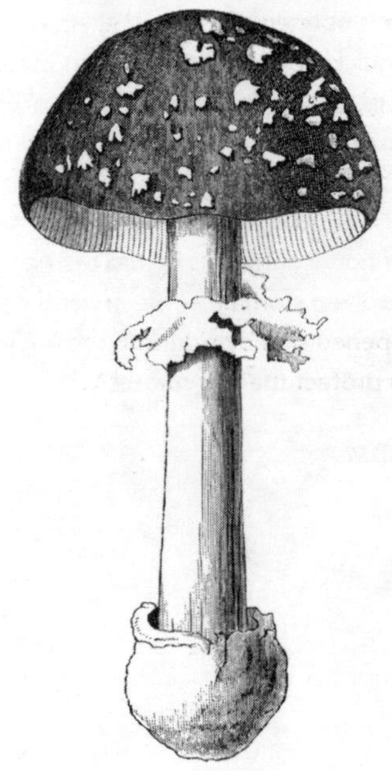

Stamen

Stamens are the male reproductive parts of a flowering plant. Their sole function is to produce **pollen**, which carries the genetic material of the plant in the form of a reproductive cell that divides to become sperm. The stamen is made up of several parts, the most conspicuous being the anther, which produces the pollen. But the word stamen comes from the Latin for "thread", probably referring to the thin filaments on which the anthers are borne. Even within a single **flower**, plants can produce numerous stamens. The saguaro cactus (*Carnegiea gigantea*) was once counted to have produced more than 3,000 stamens in a single flower.

Stomata

Stomata — from the Greek *stoma*, meaning "mouth" – are openings, usually located on the underside of a **leaf**, which let a plant control the amount of gas passing in and out of its cells. Each stoma is opened and closed using two guard cells, which can quite quickly become rigid or soft in response to the needs of the plant. Opening the stomata allows gases to pass in for **photosynthesis**, but also risks water loss, as the air inside plants is full of water vapour. Plants that live in very dry **climates** must try to keep hydrated by opening stomata during the cooler hours only. This in turn restricts their ability to grow. Plants evolved stomata as life on land began, a process that happened side by side with the adoption of waxy **cuticles** to protect the leaf surfaces.

Symbiosis

Symbiosis, from the Greek for "living together", describes a process by which two separate organisms have evolved to live in a persisting relationship. **Legumes**, for example, have nodules on their **roots** created by rhizobia infection. These nitrogen-fixing bacteria help the plant by taking in nitrogen from the air and making it biologically available, in return for compounds they need themselves. Symbiosis can be obligatory, where at least one of the symbionts is dependent on the other to survive, or it can be facultative, where it simply improves the chances of survival. Much recent botanical study has focused on how symbionts become involved in "arms races", as the hosted organism struggles to evolve ways of "cheating", while the hosts must learn to detect such rule-breakers.

Tuber

From the Latin for "swelling" or "hump", a tuber is an underground swollen stem or root produced by some plants which acts as a store of water and **nutrients**. This storage capacity enables plants that produce tubers to become dormant during winter, or periods of drought, but retain all the water and nutrients they need to begin growth when conditions improve.

Tubers are solid, and often rounded in shape. Undifferentiated inside (which means that they look the same all the way through) they have growing points – or eyes – on their upper side and will produce roots on the lower side. Perhaps the best-known tubers are edible; potatoes are an example of a stem tuber, and sweet potatoes are root tubers.

Vascular

This is the group of complex plants which contain systems of specialist cells to transport **nutrients**, carbohydrates and water around the plant. There are two systems – the xylem, for water and minerals; and the phloem, for the products of **photosynthesis** – and both act like the utilities core of a tall building, bundling services to each part of the plant together. Movement of water and useful dissolved compounds takes place as vapour evaporates from the **stomata** and is replaced from the **roots**, a neat solution that saves the plant considerable energy. Sugars produced in the **leaves** from light processes must move in the other direction, downward to the roots. The vascular plants include the horsetails, the **ferns**, the gymnosperms (or **conifers**) and the last group to evolve, the angiosperms or flowering plants.

Wood

Wood is a hard substance produced by some plants. Within all woody plants is a ring of cells – the vascular cambium – which forms new tissue, expanding the plant outwards, increasing its girth. Everything on the inside of this ring is called wood. There are two types of wood. Sapwood is the name for the youngest growth – newly formed tissue that conducts water from the roots to the rest of the plant. Over time, however, sapwood accrues waste products – oils, **tannins** and **resins**, for instance. It stops conducting water and becomes what is called heartwood. The cells which make up heartwood are technically dead, but wood cell walls contain a substance called lignin, which keeps the cells rigid even after they have died. So, both sapwood and heartwood fulfil another function of wood; they are the structural support for the large forms that woody plants have evolved (see **Trees**, page 89).

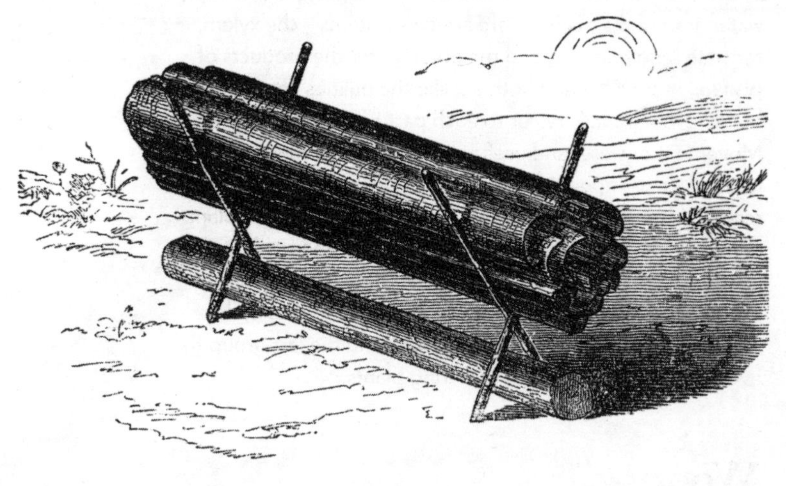

Growing

Agriculture

Agriculture is the raising of crops or animals to produce food and commodities – it is also known as farming. It is a broad term which covers varied approaches to cultivating many kinds of plants and rearing many breeds of animal, so it is best understood as a signifier of industry, separated from similar activities such as vegetable growing or keeping chickens by its scale and practice. The word is borrowed from Latin and is a combination of the words which meant "field" (*ager*) and "growing" (*cultura*). But food production in particular has expanded from the field to include a whole host of alternative approaches such as aquaponics, vertical farming, and agroforestry. These novel approaches are a possible solution to the greatest problem which agriculture faces: how to feed the world's growing population without causing irreparable harm to the natural environment on which we depend.

Blossom

Blossom specifically refers to the **flowers** of the stone **fruit** trees, which appear in profusion as the **weather** starts to improve in spring. To attract specific pollinators, the blossom is often shades of pink to white, covering almost the entire **tree**, and carrying an attractive scent. The preferred pollinators are insects, particularly bees, and blossom provides the first huge boost of the year for hives. This indicates an adaptive co-evolution, as the trees get to make almost exclusive use of these popular pollinators at their specific flowering time. Blossom has come to have a particular ethnobotanical meaning for humans, and its appearance is celebrated in many cultures for its association with the warming year. None of these are more passionate than the Japanese blossom festivals, known as *Hanami*, for which special weather forecasts are produced to allow for proper planning of the viewing to take place.

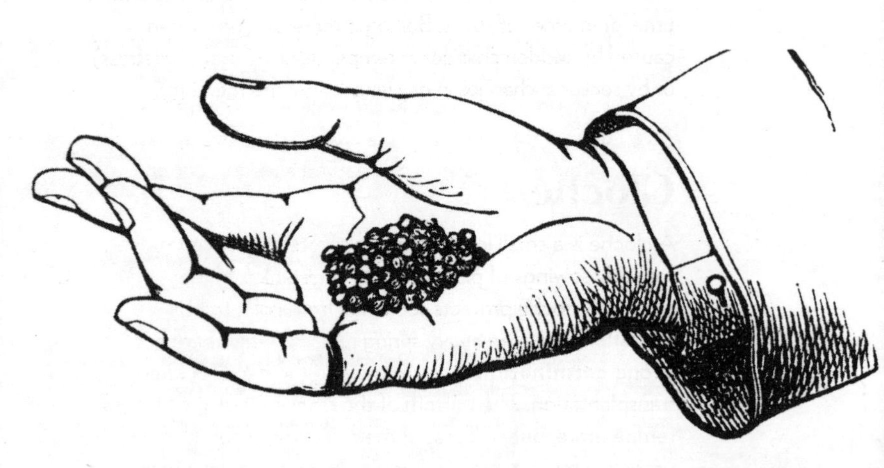

Bolting

Bolting is an informal word used to describe the premature (and unwanted) flowering of a plant. Flowering is a natural and essential part of a plant's life, but in a **garden** context it is not necessarily desired. For instance, the onset of flowering in leafy vegetables and **annual** herbs reduces production of edible foliage and changes the taste of the **leaves** so that they become bitter. The term "bolted" is applied when flowering goes against the desires of the grower like this.

Plants have evolved to produce **flowers** either at the optimum time, or in times of stress. Bolting is therefore most often caused by sudden changes in temperature, by drought (stress) or by seasonal changes in day length (optimum timing).

Cloche

A cloche is a small low transparent protective cover for early sowings of plants, mainly vegetables, in the **soil**. The cloche protects seedlings from cold, frost, predation and even heavy spring rain, and encourages strong **germination** *in situ*, removing the need for later transplantation. The warmth of the cloche raises the air temperature, but also that of the soil underneath it, to provide a much more sympathetic growing environment, advancing crop times by some weeks. Originally cloches were made of **wood** and glass, but these days lighter transparent plastic is much more widely used, though arguably less sustainably. Cloches specifically designed to prevent predation, especially by birds, may have more open coverings, such as mesh, and all will require ventilation to avoid disease caused by opportunistic micro-organisms.

Compost

Compost is decomposed **organic** matter used to improve **soil**. The word comes from the Latin *compositus,* meaning "put together". And this is the basis of compost: natural waste materials are mixed together and left to break down. The best compost is a mix of different materials, some woody (containing carbon) and some green (containing nitrogen).

In the wild, fallen **leaves** and **fruit**, dead plants and **trees** will all rot naturally. The act of composting – bringing these materials together, making sure they are kept moist, and turning the mixture to introduce air – simply ensures that conditions are perfect for the micro-organisms which do the breaking down, and so speeds up the process.

Coppicing

Coppicing is the practice of regularly cutting trees down very close to the ground. When **pruned** hard like this, certain **species** such as hazel, alder and willow will readily regrow from the base, producing stems that are straighter, more uniform and far more numerous than the tree's natural growth. These were originally used as a reliably consistent source of fuel but also found a use in fencing, construction, and – in the case of the young flexible growth – rural crafts such as basketry. Coppicing is also practised on ornamental plants. Species like dogwood and willow can be coppiced annually, which encourages them to produce a profusion of brightly coloured young stems that persist through the following winter.

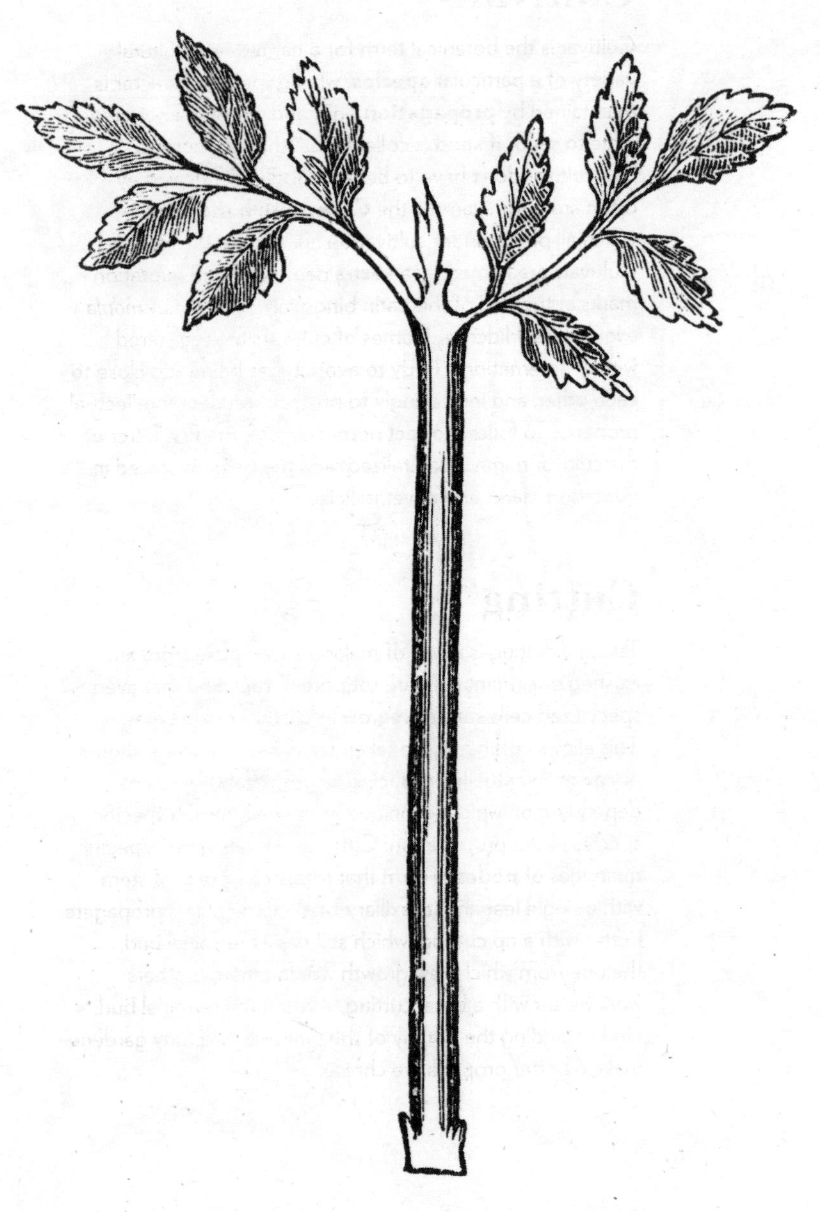

Cultivar

Cultivar is the botanical term for a named horticultural variety of a particular **species**, whose special character is maintained by **propagation** only, and which will not grow "true to seed" if **seed** is collected. Although members of a cultivar don't have to be genetically identical, they often are – for example the **Cavendish banana**, of which all plants under cultivation are genetically identical. Cultivars are named by an extra name in single quotation marks at the end of the Latin binomial – hence, *Lavandula angustifolia* 'Hidcote'. Names of cultivars are registered with an international body to avoid types being too close to each other, and increasingly to protect breeders' intellectual property. To follow correct nomenclature, the first letter of the cultivar name is capitalized, and the name is placed in quotation marks and never italicized.

Cutting

Taking a cutting is a way of making a new plant from an existing one. Plant cells are totipotent, meaning that even specialized cells can still regrow into other tissue types. This allows cuttings to be taken from stems, **leaves**, shoots or **roots**, eventually producing a completely new plant, depending on which technique works best for the specific species under propagation. Cuttings are taken with specific quantities of **nodes**, a term that refers to a piece of stem with a single leaf and its axillary **bud**. Some plants propagate better with a tip cutting, which still has its terminal bud, the one from which new growth would emerge; others work better with a basal cutting, without the terminal bud. Understanding the botany of the plant will help any gardener make a better propagative choice.

Etiolation

Etiolation is the word for how plants grow in the absence of light. Plants can continue to grow for some time without sunlight, as they push through **soil** as a germinating seeding, or from under detritus. Biologically they will grow faster than they would in strong sunlight, because finding a light source is so important. The growing tip of a plant is particularly adapted to seek even low quantities of sunlight, producing **hormones** to keep the apical tip growing until it does; this quickly gives the plant a spindly elongated shape. The lack of chlorophyll, not made until light is found, also gives the plant a pale yellow appearance. As soon as a seedling or plant reaches the light, a set of changes take place, as **leaves** and chlorophyll are produced and the rapid searching growth phase ends.

Fertilizer

A fertilizer is a substance that provides plant **nutrients**. Plants produce their own energy through **photosynthesis**. But for healthy growth they require some elements that they can get only from their surroundings: essential building blocks which they use to construct things like chlorophyll. A fertilizer is a means to ensure that these nutrients are present for the plant to use, and can be applied to the soil or to the plant itself.

A fertilizer is not a cure-all. Many nutrient deficiencies result from pH imbalances or the unavailability of water, making it difficult for the plant to take up nutrients. And excess nutrients resulting from the application of fertilizers can have unintended consequences, for both plants and the wider environment.

Forage

Foraging describes the way that grazing animals seek their plant-based food, and the way in which hunter-gatherers also did so before the coming of **agriculture**. In the past it was believed that prehistoric people went from hunter-gathering to a settled life, but it now seems clear that a mixture of both continued in most societies after crop-growing was introduced. Foraging has recently come back into fashion, although enthusiasm for finding one's own food could be dated back to Richard Maybey's *Food for Free* of 1972, which contained inviting recipes such as nettle soup. The law in the UK offers a clear definition of foraging: collecting something wild on any land, to eat and not to sell, is not stealing. However, keeping apples that have fallen into your **garden** from a neighbour's **tree**, without first offering them to the neighbour, is illegal.

Forcing

Forcing means to encourage a plant to early maturity. Plants grow according to the rhythm of the seasons. But there are ways to force a plant to come into growth or to fruit at a different time than it would naturally do. Temperature has a particular effect on plant growth cycles, so this generally involves creating a microclimate around the plant, or moving it to a warmer environment entirely. It is distinct from permanent indoor growing, however, because the plant still undergoes a period of cold. Rhubarb is a commonly forced plant. By covering the **crown** or bringing plants indoors, a crop can be harvested weeks ahead of plants out in the open. The lack of light also results in a sweeter, pinker stem, which has become a particularly sought-after seasonal food.

Forest garden

Forest gardening is a style of constructing a vegetable growing area which takes into account the natural ecology of vegetable plants, **layering** them in a way that suggests a naturally occurring **ecosystem**. The design is aimed at making a more sustainable method of food growing, and to reduce maintenance, with **trees** providing gentle shade above, and layers of productive plants below to reflect the needs of each. Forest gardening also uses companion **species**, to reduce predation by **pests**, and nitrogen fixers to increase **soil** productivity. It was explicitly conceived in response to **climate change**, with the goal of using water efficiently. It may also create other useful by-products besides food, such as canes for staking plants, fibres for tying, **wood** for fuel, **compost** for **mulch**, and honey.

Germination

Germination is the emergence of a seedling from a **seed**. A seed contains a dormant plant embryo. When conditions are right, this embryo starts using stores of energy also contained within the seed to produce new cells. It sprouts a **root**, which anchors it and allows it to take up its own water and **nutrients**. It then pushes out **leaves** into the light, where they can begin to photosynthesize and provide the new plant with the energy it needs to establish. This is germination. Germination only occurs under the right conditions. Water is key: a seed must rehydrate before it can germinate. But temperature, light levels and oxygen availability are also important.

Grafting

Grafting is the technique of attaching living pieces of two different plants together so that they function as one. It is perhaps comparable to a limb transplant in a human. It is most commonly used as a method of **propagation**: a piece of the plant to be reproduced (the scion) is attached to the root or stem of another plant (the rootstock) which then acts as the scion's root system. The main advantage of this approach is that the process confers the properties of the rootstock onto the scion. For instance, an apple variety that might grow too big for an allotment can be grafted onto a naturally smaller species and will stay small. Grafting works only with closely related species – plants from the same **genus** are most likely to succeed.

Growing medium

Growing medium is the physical material in which a plant is rooted. Plants will grow best in mediums that suit their particular needs. These needs may change greatly over their lifetimes, so growing mediums should be adjusted accordingly. **Seeds** need to be specially protected from damp and potential mould as they germinate, but also need delicate and constant moisture – **germination** can be disturbed or ruined by too much wetness. For that reason, fine gritty mediums are often best at this stage. Later, plants need more organic matter for **roots** to solidly anchor, and to provide mature levels of **nutrients**. In the past, peat has been seen as a suitable growing medium, but today we better understand the environmental impact of harvesting peat, and horticulture has found good alternatives such as coconut coir.

Humidity

Humidity is a measure of the amount of water vapour in the air, and is one of the central challenges faced by plants: their survival in a particular environment will depend on whether the humidity is within their range. If humidity is too high and air circulation not great enough, plants will rot, unable to make gaseous exchange for **photosynthesis**. Low humidity presents equally significant difficulties: when it is very dry, plants must shut their **stomata** to avoid water loss, slowing their growth. Houseplants in particular often need higher levels of humidity than most of them experience in average homes. There are lots of ways to increase humidity for houseplants, but most also increase their risk of moulds and bacteria, plus **pests** like fungus gnats.

Hybrid

A hybrid, from the Latin for "mongrel", is botanically the offspring of one plant **species** cross-fertilized by another. The first generation of crosses are known as F1, and they are valued by gardeners for being very consistent in their character, reflected in premium nursery prices. The F2 generation is more likely to be assorted in size, shape and colour. Hybrids can display hybrid vigour, where a cross is stronger in growth and more robust than either of its parents. Hybridization is a central process in the production of new varieties for horticulture, and has allowed the wide range of garden plants we see today, whose names are denoted by a multiplication sign between the two Latin names. However, it cannot go beyond the limits of the genome: despite the industry's best efforts, a blue rose has not yet been bred.

Hydroponics

Hydroponics, from the Greek *hydro-*, "of water", and *ponos*, meaning "work", is the art of growing plants in a setting without **soil**. Instead, nutrients dissolved in water form a liquid growing medium for plants, often based in a non-reactive substrate such as perlite or coconut coir. The challenge is to ensure plants are able to draw the minerals they need from the liquid, without preventing the roots from their necessary gas exchange, in particular taking oxygen from the air. Hydroponics is a very efficient method of growing in relation to water usage, and is often used to maximize growth, especially of fast-cropping salad **leaves**, herbs and other small crops, often in inclement **climates** or regions where there is little natural sunlight during certain times of year.

Irrigation

To irrigate is to supply plants with water. Water is the most crucial thing that a plant needs to survive, but the environment does not always provide enough. In the wild a lack of water would probably result in the death of a plant so, in cultivation, we intervene when natural water availability is low, using irrigation to ensure that there is enough. This is particularly necessary if a plant has been taken out of its natural **habitat** or is being grown in a pot.

Humans have known about the importance of water for healthy plants for a very long time. In what is now Iran, there is evidence dating back more than 6,000 years of canals used to divert river water for crops. But irrigation can take many forms; watering cans, hosepipes, and large-scale agricultural sprinklers are all modern methods of irrigation.

Layering

Layering is a **propagation** technique. In certain woody plants, branches are bent down to the ground and, when partially covered with **soil**, will produce **roots**. This makes the branch self-sufficient and it can then be separated from the original plant as a new individual. This happens naturally in many **species** that droop to the ground such as brambles, but the technique can also be used on other plants, including **climbers** like jasmine and shrubs like magnolia.

An alternative approach is called air layering. Rather than being bent down to the soil, branches are wounded and kept moist with a wrapping of sphagnum moss or a similar material until they develop roots *in situ*.

Loam

The word loam describes a certain kind of soil. Geologically, the word refers to a soil that contains a mixture of particle sizes; varying quantities of sand, silt and clay. There are sandy loams, silt loams and clay loams, and everything in between.

But this is the most technical definition. Because a combination of different particle sizes results in a good soil for growing – one that holds on to **nutrients** and water but doesn't get waterlogged – the term has come to be used in horticulture and **agriculture** to refer to any soil with these qualities. Generally this means a soil that has plenty of humus, or rotted **organic** matter, in it.

Micropropagation

The science of growing a whole new plant from a very tiny quantity of cell tissue, micropropagation is particularly important in a **conservation** context, where remaining wild individuals of an endangered species may number in the hundreds or even tens. Micropropagation is also a central process in commercial growing, especially for plants like orchids whose **seeds** are extremely small and germinate most effectively with the correct **mycorrhizal** fungus. Pioneered in the late 1950s at Cornell, by Frederick Campion Stewart, micropropagation takes place in a highly sterile laboratory setting. Although a lab process, micropropagation is a highly artful area of horticulture and may require many special skills to bring to fruition: at Kew it has been used effectively to send endangered and even extinct species back to the wild, including *Trochetiopsis ebenus,* the St Helena ebony, of which only two wild plants then survived.

Mulch

A mulch is any material purposefully spread on top of the **soil**. Mulches can confer numerous benefits. They provide insulation and protection from erosion, and help with moisture retention and **weed** suppression. Depending on the intended purpose, many different materials can be used as a mulch, including woodchip, gravel and even black plastic sheeting. The word stems from the German *mölsch*, which means soft and rotten and, indeed, **organic** matter is perhaps the most commonly used material. Annual mulching with organic matter is the basis of "no-dig" growing methods, and experiments have shown that it can produce better results than more traditional methods of incorporating organic matter into the soil every year.

Nutrients

From the Latin word for "nourish", nutrients are the chemical elements required by plants for healthy growth. Technically this includes elements such as the carbon and hydrogen needed for **photosynthesis**, the oxygen required for respiration. But these are obtained from air and water, which must be regularly replenished for the plant to even stand a chance of survival, so the term nutrients is often used to mean just those elements obtained from the **root** environment. The three most important of these are nitrogen, phosphorus, and potassium (labelled on packets of **fertilizer** as their chemical symbols, respectively NPK), but there are numerous others used by plants in varying amounts, including sulphur, calcium, magnesium, iron, and copper. These are the building blocks from which a plant constructs its cells and their absence manifests as various symptoms, from yellowing **leaves** to stunted growth, depending on the nutrient that is lacking.

Organic

This term, meaning "relating to living things", has come in the last 30 years to refer specifically to methods of growing that work with and not against the life cycle of natural **ecosystems**. Organic horticulture avoids the majority of chemical methods of pest control, preferring physical traps such as sticky sheets, or using natural predators such as nematodes. It also chooses nitrogen fertilizers that arise from within the natural cycle of life, such as manure, rather than those made by the chemical industry. It may also include "no-dig" methods, which aim to preserve the structural integrity of the soil, allowing better **mycorrhizal** systems to develop and also the retention of more surfaces for the storage of **nutrients** and for roots to achieve gas exchange.

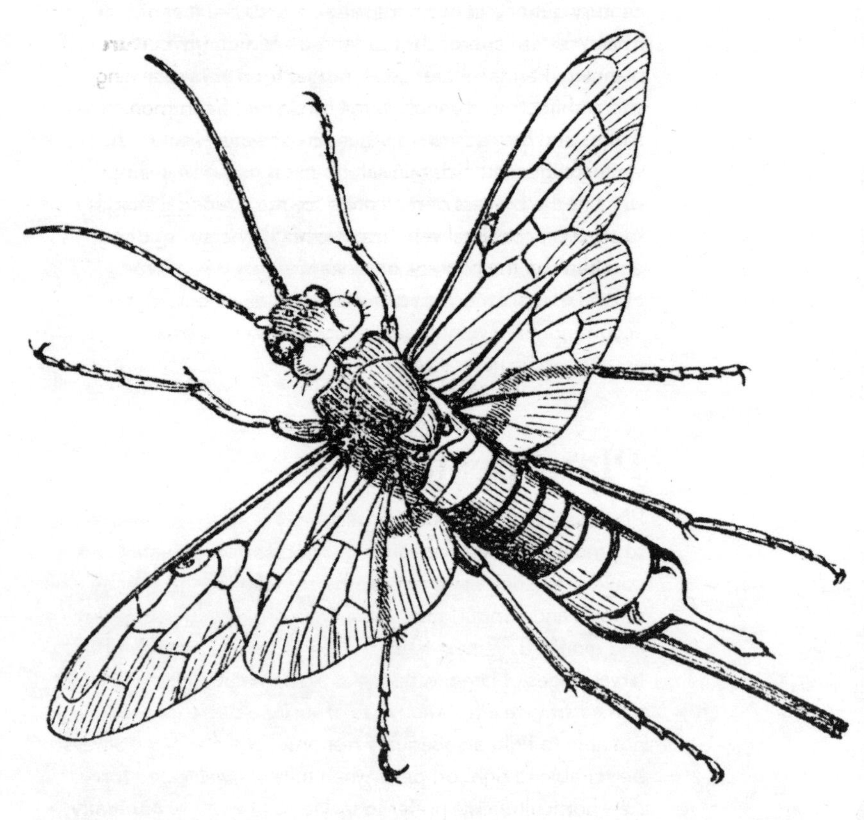

Permaculture

Permaculture is a portmanteau term originally deriving from "permanent agriculture" and describes a "whole systems" view of land management. It developed in the late twentieth century out of a concern for the non-sustainability of most Western approaches to settlement and **agriculture** planning. Permaculture's remit ranges from town planning and architecture, through to the rewilding of common spaces and the related management of water resources via techniques such as rainwater harvesting and remiring, which is the process of recreating former drained wetlands to absorb occasional very heavy rainfall. There is a strong emphasis on the concept of resilience within a system of land use, allowing for **climate** and other shocks to be accommodated without jeopardizing either humans or aspects of the natural world.

Plant pests

Plants are stationary, and consequently are very vulnerable to predator attack. The range of **species** that arguably could prove pests to plants is almost unlimited, from slime moulds and **fungi** through to large mammals such as deer and giraffe. The most familiar pests for plants are often larval stages of larger arthropods, such as cutworms and leaf miners. Severe infestations can often be a sign that a plant is not able to fight an adequate response, and plants are always better able to fight off pests when they are well cared for. Kew horticulturalists prefer to tackle plant pests **organically**, which in many cases means setting yellow sticky traps to physically reduce predator numbers.

Pollarding

Pollarding is the practice of regularly cutting off the upper trunk and branches of a **tree**. This encourages lots of thinner growth which, like coppice wood, would traditionally have been used for weaving and basketry, or left for slightly longer to use as fence posts. But, unlike **coppicing**, pollarding leaves the bottom of the trunk intact – the term comes from the word *poll*, which meant to cut the hair in Middle English. Only certain **species** will react well to pollarding. Today, the practice is primarily used to control the size of trees so that they do not outgrow their space, particularly in cities where large trees can impact buildings, cast shade, and disrupt overhead services. It is most often seen on city trees like the London plane and the common lime.

Pollination

Pollination is the transfer of **pollen** from **stamen** to stigma. It is the most vital step in a flowering plant's sexual reproduction: the way in which one plant's male sperm cells are delivered to the female egg cell in another plant's ovule for fertilization. Some plants rely on currents of wind or water to carry their pollen. Many others have evolved symbiotic relationships with insects or animals; tempting pollinators into their **flowers** with bright colours and the promise of nutritious **nectar**, they deposit pollen on their visitors who then spread it to other plants. In some flowering plants pollen transferred from the stamen to the stigma within the same flower will result in fertilization, but hybridization with another plant is far more advantageous, so many others have evolved mechanisms which mean they are self-incompatible.

Propagation

Propagation is the general term for the creation of new plants, ranging from sowing **seed** to taking **cuttings**. It is divided into two main types: sexual and asexual. Sexual propagation is the **germination** of seeds and **spores** into new plants, after recombination by mixing genetically with the cell line of another individual. Asexual propagation is the creation of clones from plant tissue, whether by taking cuttings, rooting runners, **layering** or **grafting**, or at the cell level. Asexual reproduction also includes growing seeds where they have been produced asexually, by apomixis. This is the process by which some plants have evolved the ability to make seeds in which the embryo will develop as a clone of its mother plant, and it is widely known in plant families such as maize and wheat. It may be opportunistic in particular circumstances.

Pruning

Pruning is the considered removal of material from a plant. It is not necessarily an essential act for a plant's survival; plants live in the wild without anyone to trim their branches (though the removal of diseased material perhaps gives them a better chance). But pruning does allow the grower to exercise some control over their plants. This can be as simple as regulating size or making the shape of a plant more attractive. But pruning is also used to encourage certain kinds of growth. The annual pruning of rose bushes and apple trees, for instance, promotes improved flowering and fruiting respectively. The term is a borrowing from Old French, where the word was *proignier*.

Scarification

Many **seeds** have evolved to have very hard outer coatings that prevent the seed accidentally germinating before appropriate. In natural conditions, water, temperature, air and even fire would eventually work on the seed to ensure **germination**. In order to coax these kinds of seeds artificially, the hard seed coat is scarified, meaning that it is treated with methods to mimic those that occur in the wild, involving water, acid, heat, temperature or by physically scratching some of the coating off. Many **Mediterranean**-type scrub **species** require exposure to heat and or smoke to germinate optimally – for example, the stone pine cone is more likely to germinate after a single peak of over 100 °C, such as a fire would create, than in a normal summer of many days at 40 °C.

Senescence

Senescence is the natural process of decline, which happens as a plant nears the end of its life. From the Latin *senescere* (to grow old), senescence is distinct from the decline seen when a plant is affected by external factors such as disease, **pests** or lack of resources. It is caused by deterioration at a cellular level, which accumulates over an organism's lifetime.

Senescence is an unavoidable and irreversible process, and will eventually result in the natural death of a plant. Woody plants in particular, however, are amongst the longest lived of all organisms and can avoid senescence for a very long time. There is a living bristlecone pine in California which is thought to be over 5,000 years old.

Shelter belt

While traditional forestry often espoused single-species plantations, today there is more understanding of the value of mixed planting in providing more biodiverse **habitats**, as well as more varied physical shelter from wind and heavy **weather**. Experiments in the very windswept Outer Hebrides, for example, have shown that although commercial wood-growing is not possible, human-made shelter belts create conditions in which self-sown natives can begin to establish. The Royal Botanic Garden Edinburgh makes use of a series of shelter belts, dominated by pines and hollies, to protect a range of plants that otherwise might not flourish. A shelter belt mimics the way in which seedling **trees** would grow in the wild, in small cleared areas within mostly forested ground, and thus is a useful and sustainable forestry technique.

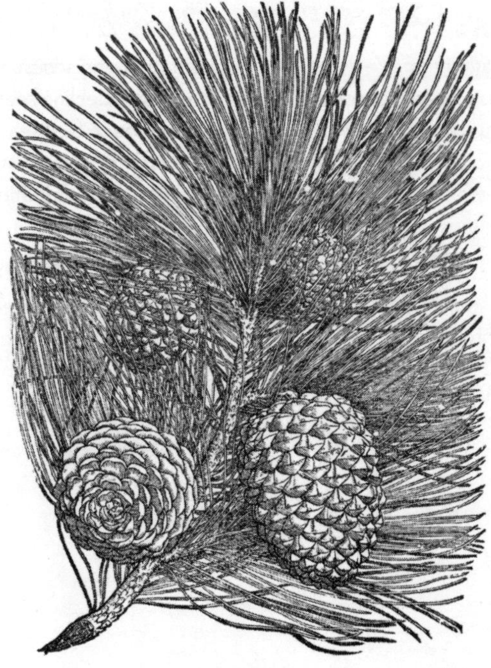

Terrarium

A terrarium (plural: terraria) is a glass enclosure in which live plants are kept. In modern usage a terrarium is also understood as a decorative way to keep plants indoors; a living landscape of plants in miniature contained within a glass display case. The glass sides allow in light and heat, but also mean that moisture is retained and heat loss is limited. Even in an open terrarium (such as a jar, or a bowl without a lid) a small microclimate is created which provides perfect conditions for certain plants to thrive. Within a sealed terrarium, air, water and nutrient cycles can be created in miniature and – in theory at least – the plants can essentially look after themselves. One terrarium, created by an enthusiast, David Latimer, hasn't been opened since 1972.

Vegetative reproduction

Vegetative reproduction is the asexual process by which a plant can create an identical clone of itself. It is a strategy used by almost all plants. A well-known example is the spider plant, which produces tiny plants on graceful stems, but in other plants ranges from producing bulblets around the **bulb** underground to sending out runners, aerial stems that will root themselves when they find an appropriate spot. It allows a successfully adapted individual to make more copies of itself when conditions are adaptive, increasing the plant's population and allowing them to grow rapidly to size using access to the mature plant's nutrient and fluid systems. Vegetative reproduction is facilitated by a plant's ability to make adventitious **roots** – roots that grow downwards to find earth. The major drawback of vegetative reproduction is the lack of genetic diversity associated with non-sexual reproduction.

Plant Types

Alpines

An alpine is any plant native to high mountain slopes. Although the word means "of the alps", it is applied to plants from all the mountainous regions of the world, not just this specific range. True alpines grow above the treeline, the height on a mountain at which conditions become too hostile for **trees** to survive. So, alpines must be adapted to difficult growing conditions. They can survive cold temperatures and high amounts of UV light; the majority are **perennial** with hardy **root** systems that store food through the winter, allowing them to flower quickly and take advantage of the short season of more favourable **weather**; they are low-growing to escape the worst of the wind; and they thrive in thin, nutrient-poor **soils**. Such adaptations have allowed alpine plants to colonize the most unlikely of places, including the slopes of Mount Everest, up to 6,400 metres above sea level.

Annuals

An annual is a plant that completes its entire life cycle within one growing period. Annual – from the Latin for a year, *annus* – refers to the timeframe in which this generally happens, but nature does not necessarily line up exactly with the calendar year. It is better to think of it in terms of growing periods; an annual plant will germinate, reach maturity, reproduce and die without ever entering a period of dormancy. And this is a natural death, not one caused by disease or pest attack. Because no energy is set aside to help the plant overwinter, this energy can be used to produce a huge amount of **seed**. Many common **weeds** are annuals, and the success of this strategy is apparent in how easily they proliferate.

Aquatic

An aquatic plant is one that is adapted to life in the water. Although water is a vital necessity, plants also require other things to survive: carbon dioxide and sunlight to **photosynthesize**, oxygen to respire. These necessities are more difficult to access in an aquatic environment, so plants that grow permanently in the water have had to adapt. Some of these specialists are emergent; rooted on the bottom, their stems grow up through the water to float **leaves** on the surface or hold them out in the air. Others float untethered, with their **roots** trailing beneath them. A few are submergent, existing entirely underwater. Seagrasses are the only flowering plants to live fully submerged in the sea (seaweeds are **algae**) and are an important example of this group, increasingly understood as a vital marine **ecosystem**, and valued for their role as a carbon sink.

Biennials

A biennial is a plant that completes its life cycle in two periods of growth. In the first period the plant produces only roots, stems, and leaves. In the second period the plant produces **flowers** and subsequently **seeds**, after which it dies. There is often a period of dormancy between these two periods of growth when parts above ground may even die back completely before regrowing.

Generally this cycle will take place over two years – hence the word's origin in the combining of the Latin *bi* (two) and *annus* (year). But biennials do not strictly adhere to this timeframe. Depending on conditions, some may flower in the same year that they germinate, while others may take more than two years to reach maturity. Practically, it is perhaps better to remember that they are monocarpic, which means that they flower only once before dying. Examples of biennial plants include hollyhocks and foxgloves.

Bonsai

Bonsai is the art of growing dwarfed **trees** and shrubs in containers. In Japanese, *bon* means "basin" or "bowl", and *sai* means "planting". Contrary to popular belief, bonsai are not genetically dwarfed. It is the restriction of the roots in the pot and the careful **pruning** of the branches which keeps the tree small. Traditional bonsai aims to recreate the natural form and shape of a fully grown tree in miniature, so the pruning is highly selective and branches are trained using wire to mimic the natural form of much larger specimens. The skill involved in creating a beautiful bonsai, and the sheer amount of time and effort it takes, means that the best specimens can become very expensive. In 2011, a choice pine – thought to be centuries old – sold for $1.3 million.

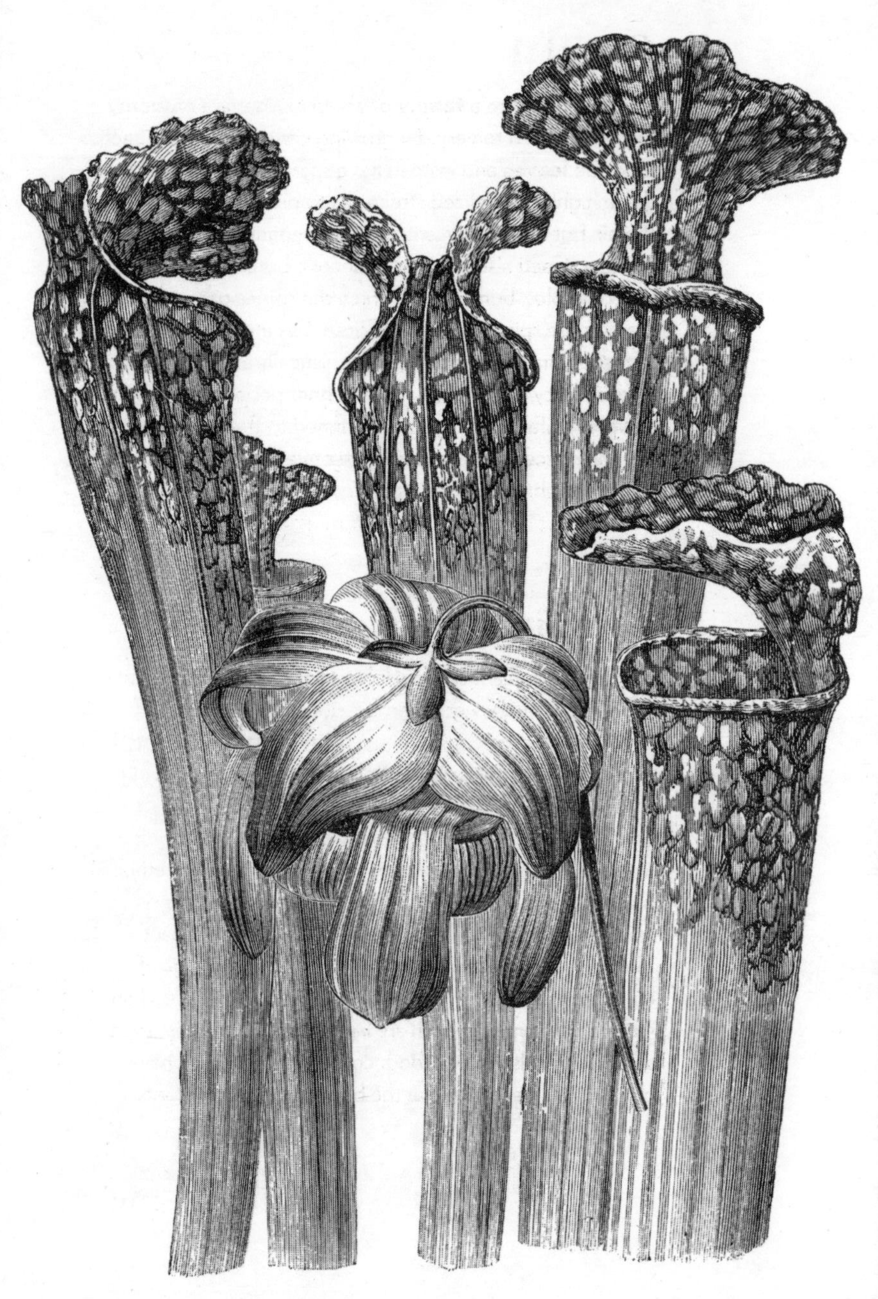

Cacti

The true cacti are a **family** of about 127 genera of thorny plants, adapted to very dry growing conditions. Most cacti lack true **leaves** and instead have spines. They evolved these highly specialized structures to prevent water loss via the air, but also to prevent animals predating their water content. Cacti also possess distinctive botanical features called areoles, bumps from which the spines grow, allowing many more spines per surface area. Cacti **photosynthesize** in their stems and flower opportunistically after long periods of dormancy, making the most of brief periods of adequate **desert** rainfall. The cacti are confined to the New World, with one exception, the **epiphytic** mistletoe cactus *Rhipsalis baccifera*, which has managed to colonize Africa and Sri Lanka, most likely on the feet of migratory birds.

Carnivorous

A carnivorous plant is one adapted to grow in low nutrient conditions, usually wet anaerobic sites such as **bogs** and **swamps** or at the edges of bodies of water, where plants' best adaptive strategy is to trap animals to provide the missing **nutrients**. The animals are lured in by a variety of means. Some carnivorous plants use scent or strong colouration to attract prey, then use one of five possible trapping methods: pitfall, adhesive, snap, snare and suction. The strategy is so effective that it has evolved in several different plant families independently. There are about 650 **species** of carnivorous plant known to science, most of which feed on insects, but some of which are large enough to digest even small mammals. At Kew, both cockroaches and rats have occasionally been found at the bottom of pitcher plants.

Cereals

A cereal is any plant of the **grass** family (Poaceae) whose **seeds** are eaten by humans. Named for Ceres, the Roman goddess of **agriculture**, cereals are some of the most cultivated plants on Earth. They are a rich source of energy and **nutrients** and can be easily stored and transported. Combined with their ease of cultivation, this has made them a staple crop throughout history. With evidence of their cultivation stretching back 9,000 years, they have played a key role in all the great civilizations of the world. They remain essential to this day; rice, wheat and corn are all cereal crops, and it has been estimated that more than 4 billion people rely on these three crops as their main diet.

Climbers

Climbers are plants that make use of other plants or vertical surfaces to extend their growing area, and to escape ground-based grazers, while expending relatively few resources in a trunk or other vertical structure. There are climbers in many different plant families because it is an adaptive evolutionary strategy, rather than a related group of plants. Climbers use a wide range of means, from twining, where the main stem curves around, to tendrils, in the case of the pea and bean **family**, to suckers, as in plants such as ivy. Climbers must have high tensile strength and flexibility in their stems to use both natural and human-made structures. Some tendrils are specially adapted **leaves** or even parts of the **flower**, such as the thorny "hooks" of *Bougainvillea*.

Conifers

Conifers are a type of woody plant. The word comes from the Latin and means "cone-bearing", and this is a main defining feature of the group. Rather than producing flowers, conifers hold their reproductive organs and subsequently their seeds in protective structures called **cones**. But **cycads** also produce cones, so conifers are additionally distinguished by their habit and foliage. The majority are **evergreen trees** or shrubs – of more than 600 accepted conifer **species**, fewer than 20 are **deciduous**. And almost all species have evolved specially adapted leaves – needles or scales – the shape of which helps to reduce water loss. Conifers dominate in colder regions – in northern latitudes, and on mountain slopes – and adaptations like this help them to survive the winter when water is unavailable, frozen in the ground.

Crop wild relatives

As **agriculture** faces new challenges due to our changing **climate**, many crops are threatened by increasing insecurity. Food crops need to be grown in altered conditions and their limited ability to adapt puts the harvest at risk. For this reason, scientists are taking a growing interest in the wild close relatives of Earth's most important food crops, including rice, **coffee**, chickpeas and **bananas**. Plant scientists are studying and conserving **seeds** of these crop wild relatives because they have more resilience than their cultivated cousins, surviving harsher conditions, including the ability to cope with drought, flooding and salinity. By preserving and banking their seeds, breeders will have access to new genetic material to develop new resilient crop varieties. Many of these wild relatives are disappearing from their natural **habitats** too, due to pressures such as deforestation, intensive agriculture and urbanization. Kew is working closely with Ethiopian coffee growers, because over half of coffee's wild relatives are in danger of **extinction**, threatening their rich and useful genetic inheritance.

Cryptogam

Cryptogam, from the Greek *kruptos*, "hidden", and *gamos*, "marriage", indicates hidden reproduction, for the cryptogams are the organisms that reproduce without making seeds. Instead, this group of plants makes **spores**: it includes **algae**, **moss**, **lichen** and **ferns**. **Linnaeus** originally devised the class, including the plants previously listed as well as **fungi** and slime moulds, which today would be recognized as belonging to an entirely different natural kingdom. Thus today it is an informal grouping, rather than an evolutionarily-related clade. Algae for example is today seen as an informal term, comprising as it does both tiny blue-green algae right up to giant kelp, the largest algae on earth, which can grow to be 50m from base anchor to tip. Despite its appearance, giant kelp is not a plant, but a heterokont, a member of the kingdom Chromista.

Cycad

Cycads are a **family** of about three hundred **species** of cone-bearing plants, usually with trunks topped by a rosette of stiff radiating **leaves**. This gives them an appearance like a **palm**, although they are not closely related. The cycads are very long-lived plants, and some examples are known to be many hundreds of years old. Kew, for example, has one collected in the wild in 1776, which is still in a pot (admittedly, a large one); cycads are very slow-growing. Cycad **cones** are usually fertilized by specific species of beetles, and perhaps due to this and their very slow growth many species of cycad are severely threatened in the wild, though they are popular **garden** plants.

Deciduous

Deciduous plants are those which shed all their leaves on a seasonal basis. From the Classical Latin *deciduus,* which means "tending to fall", the term is generally used for shrubs and **trees** that drop their leaves but retain a woody structure. In colder **climates**, leaf drop (or abscission) happens in preparation for winter. Having no leaves removes the possibility of frost damage and reduces strain on branches from winter winds. But its primary advantage is to limit water loss when water may be unavailable, frozen in the ground. Deciduous tree species can therefore also be found in areas that experience annual drought; they lose their leaves to retain water during the dry season. The process of abscission is actively initiated and carefully regulated by plant **hormones**, and it is a visible process. The slowed production of chlorophyll (the pigment which makes leaves green) and its eventual breakdown means that other pigments gradually become visible, marking deciduous species with the bright colours we associate with autumn.

W.G.S.

Ephemeral

From the Greek *ephēmeros* which means "lasting only for a day", ephemeral describes a plant that is very short-lived. This can mean plants whose entire life cycle from **germination** to death is extremely short. Or it can mean plants that have visible leafy growth for only a very short time before dying back to re-enter dormancy.

In both cases brevity can be advantageous because it minimizes the plant's potential exposure to danger. Dormant **root** systems or dormant **seeds** are an easier form in which to ride out unfavourable conditions. This is a particularly good adaptation to hostile environments like **deserts**: here, ephemerals are well equipped to take advantage of all too brief wet spells but are more immune to the extreme drought conditions that prevail.

Evergreens

Evergreen describes those plants which retain **leaves** throughout the year. This is not to say that these plants don't ever lose leaves; all leaves have a finite lifespan. But unlike **deciduous** plants, true evergreens rarely lose all of their leaves at once, instead replacing them on rotation. This means that they can still photosynthesize on sunny winter days, and do not expend resources regrowing all their leaves in spring, possibly giving them an advantage in nutrient-poor **soils**. However, retaining their leaves means that evergreen plants have had to adapt in other ways to cope with the stresses of winter. For instance, many evergreens have leaves with thick waxy coatings which tend to have fewer **stomata** (where gas exchange takes place); such adaptations help to limit water loss.

Ferns

Ferns are a kind of plant distinguished by the way that they reproduce. They do not form **flowers** or seeds, instead dispersing their genetic material in the form of **spores**. But a spore does not directly become a new fern. It develops into what is called the gametophyte, a secondary form of the fern, which exists solely to facilitate the union of male and female sex cells. It is only when these meet and fertilization occurs that what we would recognize as a fern (which is known as the sporophyte) begins to develop. Ferns are coveted by gardeners for their intricate leaves, which are also known as fronds. And these may be where they get their name; etymologists theorize that the word fern comes from the same root as the word feather.

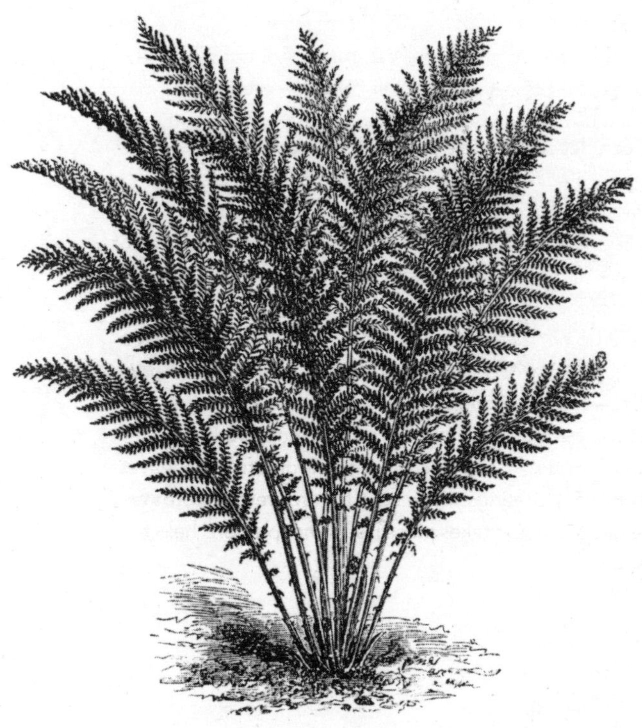

Food plants

A food plant is any plant eaten by humans as a source of sustenance. Like all other animals, we humans rely on consuming other organisms for our energy and **nutrients**: we are heterotrophs. Specifically, we are omnivores, meaning that we have evolved to eat both plants and animals. But plants make up the majority of our calorie intake and are an incredibly important source of nutrition.

We have come to think of food in quite a broad way (think of the trend in Michelin starred restaurants for foams, for example). But a food plant is probably best defined as something which provides nutritional value. It is debatable whether herbs and **spices** are food plants. We can consume them without harm, they add flavour and enjoyment, but we couldn't survive by eating them alone. A huge number of plants are safe for humans to eat. But of these, only a handful are actually consumed in any great quantity. It has been estimated that there are more than 50,000 edible plants in the world, but almost 90% of human energy intake comes from just 15 plants.

Food plants are not necessarily eaten in their entirety. It might be only the **fruit**, or the **tuber**, or the stem which is fit for consumption. In fact, many plants are poisonous in all but their edible parts. Rhubarb is a perfect example: while the stems are a delicacy, the **leaves** contain enough oxalic acid to induce vomiting, stomach cramps and even kidney failure if eaten in sufficient quantities.

Geophytes

Geophyte – from the Greek *ge-*, "earth", and *phyton,* "plant" – describes any plant that has developed the adaptive mechanism of storing food and water resources underground in a specialized storage organ. The most commonly known of these organs is probably the **bulb**, but the group also includes **corms**, tubers and **rhizomes**. These plants also protect their initial **buds** by waiting underground, which is actually the quality used botanically to distinguish them as a group. These plants manifest as a bud on an underground storage organ, waiting for a good moment to begin full growth. Geophytes can use specialized structures called contractile **roots** to move their bulb deeper, or to a different place in the **soil**.

Grasses

Grass is the **common name** for the plant **family** Poaceae. It is a large family with almost 11,000 known **species**, found all over the world in a wide range of **habitat** types. Many grass species are of great economic importance, particularly as staple food crops – wheat, rice, and corn for example.

There is huge variety in the group, but grasses are widely recognizable by their bladelike **leaves** which are held on alternate sides of hollow, jointed stems. Individual grass **flowers** are quite insignificant (all being wind-pollinated, they have no need to draw attention to themselves), but the way in which they are held in feathery **inflorescences**, and the subsequent structural **seed** heads, can be very striking. Grasses have adapted to natural pressures such as grazing and wildfires by having their growing points at the base of their leaves. This is why we are able to mow our lawns so regularly without doing any lasting damage.

Hallucinogenic

Hallucinogenic plants contain psychoactive compounds, which change perception, emotion and awareness in human beings. There are also many psychoactive **fungi**. Many of these have been valued in traditional cultures for their powerful effects, especially with respect to shamanic healing and vision practices, creating a perceived link between the living and the spirit worlds. Nowadays most hallucinogenic plants and fungi are to some extent controlled. For example, it is illegal in America to use peyote cactus, which contains high levels of the vision-inducing **alkaloid** mescaline, and the plant is also protected under **CITES**. However, there is an exception for members of the Native American Church, who are still permitted to use it in their religious ceremonies.

Halophytes

Stemming from the Greek words *álas* (salt) and *phyton* (plant), halophytes are plants that can tolerate salty conditions. Salt poses a particular problem for most plants: not only is it toxic if absorbed in large enough quantities, it can also interfere with the way that plants take up water through their **roots**. Very salty water can even dehydrate plants, drawing moisture out of their cells. The halophytes have, however, evolved mechanisms to counteract these difficulties.

Naturally, most halophytes are found in coastal areas. In **temperate** latitudes common examples include **saltmarsh** plants like the samphires (*Salicornia* species), sea lavenders (*Limonium* species) and sea beet (*Beta vulgaris* subsp. *maritima*). In more **tropical climates** the mangroves dominate – salt-tolerant **trees** that form half-submerged **forests** at the water's edge.

Healing plants

The earliest medicaments used by humans came from plants, and for many people globally herbal medicine remains an important body of knowledge and use. China and India in particular have long and substantial traditions of herbal medicine, continually practised, which have now spread in use outside the originating cultures. Herbal medicine is often perceived to be "gentler" in effects than modern pharmacology, but plants contain powerful chemical compounds to repel **pests** and invaders whose potency has made them the basis for much Western medical drug research. Plant medicines known throughout human history, and still commonly in use globally, range from aloe vera to aspirin and morphine.

Herbaceous

Although we generally think of a herb as a culinary aromatic plant like rosemary or thyme, in botany the term refers more generally to plants that don't produce any **wood**. Herbaceous plants can persist from year to year but – unlike trees, or shrubs – do not produce any permanent woody tissue. The term "herbaceous perennial" is generally used to refer only to those plants which die back to their roots in the winter months, as differentiated from **evergreen** perennials which (although technically herbaceous) retain their **leaves** year-round.

In **garden** design, "herbaceous" is a word commonly associated with the phrase herbaceous border. This simply means an area dedicated to growing these herbaceous **perennial** plants, and which will be viewed from only one side, as opposed to a bed that can be viewed from multiple directions.

Hermaphroditic

This term comes from the Greek mythological character Hermaphroditus, who combined with a nymph to make one being, of both sexes at once. In botany it means possessing both female and male reproductive organs on the same plant, so being able to produce both **pollen** and ovules. Sexual reproduction is costly, and plants have evolved strategies to avoid wasting their energy only to end up fertilizing themselves, called sexual incompatibility. Monoecious plants produce separate male and female **flowers**; the alternative strategy is to be dioecious, having male and female flowers mature at different times. The giant waterlily *Victoria amazonica* is pollinated by beetles, which are attracted by a heady perfume. The flower then shuts, trapping the insect; the next day, the waterlily's pollen becomes ripe, and brushes off on the beetle as it leaves, ensuring the plant does not fertilize itself.

Krummholz

From the German *krumm*, for "crooked" or "twisted", and *holz* meaning "**wood**", this word designates the twisted and wind-blown **trees** seen in **habitats** exposed to fierce cold and winds, such as at the treeline. Here, trees are at the edge of their ability to survive, and are significantly stunted by the struggle to grow. This distinctive kind of vegetation is almost always coniferous, such as pines and spruce. The tree gets its shape over time, by building up what is called reaction wood, where constant pressure from the prevailing wind causes the trunk and branches to grow differently than they would when guided just by gravity. The wood is correspondingly denser than normal wood from the same tree, and has a higher lignin content. Hunters in the **Arctic** have therefore often harvested this wood for longbows, because of its useful properties of strength and flexibility.

Legumes

A legume is a member of the Leguminosae, nowadays known properly as the Fabaceae, which produce a particularly adapted **fruit** not known in any other botanic **family** – a round dry fruit that usually splits along the middle into two, creating the characteristic "split pea". Their fruits are also known as pulses, when dried and high in protein, and thus are an important element of human nutrition, especially in places where meat or fish is not easily available or affordable. Another significant adaptation of the plants in this family is that most have **root** nodules containing rhizobia, **species** of nitrogen-fixing bacteria. These bacteria live in **symbiosis** with the plants, producing nitrogen for them and increasing their yields. The mutually beneficial relationship allows the legumes access to a high level of amino acids for protein synthesis, which means that legumes are among those plants with the highest levels of protein.

Medicinal plants

Because plants can neither flee from predation nor actively defend themselves physically, most of the defences they have evolved against predators are chemical. As a result, plants contain thousands of biologically active compounds, only a few of which have been fully investigated by Western medicine. Plant medicine dates back to the beginning of civilization, and was a substantial field by the time of **Dioscorides**. These days we divide medicinal plant compounds by their biochemistry into four main classes: **alkaloids** such as opium and nicotine; glycosides such as digitalis; polyphenols such as the plant oestrogens; and terpenes, which are present in essential oils. Kew estimates that about 18,000 plant species have medicinal uses, and in modern Western medicine about a quarter of all drugs prescribed derive originally from a plant compound.

Mosses

A moss is a type of simple plant. The cushion-like growths we know as moss are actually many tightly packed individuals. A single plant is extremely small, has only simple stems and **leaves**, and lacks true roots. Mosses do not flower and reproduce by **spores**.

Despite their simplicity, mosses are extremely important. They are able to colonize even bare rock, and can play a part in the process of **soil** formation. They form important relationships with other organisms which fix nitrogen, making it more available to other organisms too. In other words, they pave the way for more complex plants.

Thanks to their lack of true **vascular** tissue and inability to develop supportive tissue, mosses are generally low growing. The tallest moss in the world (*Dawsonia superba*) can grow to only 50cm (20in) high.

Palms

Palm is the **common name** for any plant belonging to the **family** Arecaceae. They are so named for the resemblance of the spread **leaves** in many **species** to an outstretched hand, hence the word's origin in the Latin for "hand", *palma*. But there is a huge variety of different types and forms, and many palms don't look like this at all.

Humans have cultivated palms for millennia and numerous palms such as the coconut and the oil palm remain important economic crops today. The oil palm in particular is a controversial crop because, although it is used in almost every kind of product imaginable, its cultivation precipitates large-scale deforestation and **habitat** loss in areas of incredible **biodiversity**.

Parasitic

A parasitic plant is one that relies on another without offering anything in terms of a symbiotic return. There are about 4,500 species of parasitic plant in total, and parasitism has evolved at least 12 times among **vascular** plants. These can be divided into plants that cannot manage without a host (an obligate parasite), and those which merely derive advantages from having one (a facultative parasite). Mistletoe is classified as an obligate hemiparasite (from the Greek *hemi-*, or half), because it can photosynthesize, especially while it is in the seedling stage, as its germinated seed **root** is slowly penetrating the outer layer of the host tree's **bark**, which may take many months. Research has shown that some parasitic plants grow almost completely within the host plant, emerging only to flower, using resources which greatly weakens the host.

Perennials

Perennial refers to plants that live for longer than two growing seasons – i.e. are not **annual** or **biennial**. From the Latin *perennis*, which is a joining together of *per* (each) and *annus* (year), perennials may die back to their **roots** in the winter but will reshoot the following spring when conditions improve.

Trees and shrubs are perennial in that they persist from year to year, but the word is most often used in relation to plants – **deciduous** or **evergreen** – that do not produce woody tissue.

Some perennials, particularly **grasses**, produce beautiful **seed** heads which persist through the winter. But these are biologically dead, which is why they can be cut back at any point without harming the plant.

Succulents

Succulents, from the Latin *succus*, meaning "juice", are an informal grouping of plants, not a botanical category. Succulents have adapted to conditions of extreme water shortage by evolving **leaves**, stems or roots that act as water storage organs. They are swollen in appearance by comparison with **herbaceous** plants, and are often shiny, due to a waxy protective outer layer that further reduces water loss. Succulents can use water slowly, and dry up almost completely before they will die, allowing them to survive dry spells that would kill most other plants. They may also have other adaptations, such as reduced numbers of **stomata**, and crassulacean acid metabolism (CAM), a biochemical pathway that allows gas exchange to take place only in the cool of night, reducing the amount of water lost by opening the leaf's pores (stomata).

Trees

A tree is a larger plant that has a long woody stem, from which branches and leaves grow outwards. Being a tree is an evolutionary strategy to reach the light and evade grazers, and involves growing a stiff woody trunk and also **vascular** tissue to carry gases, water and **nutrients** around the tree. A shrub, on the other hand, is an informal term for a small- to medium-sized woody plant. Sometimes shrubs and trees are differentiated merely by size, sometimes by the proliferation of stems which is more typical of a shrub than a tree. A 2015 study published in *Nature* estimated that the global number of trees is three trillion, with about half in **tropical** and **subtropical forests**, and the rest in higher latitudes.

Variegation

Variegation describes a change of colour to the **leaves**, veins, stems or **fruits** of plants, in a zonal pattern. The pale parts of a variegated plant usually appear thus because they lack chlorophyll, the green photosynthetic pigment. Variegation can also be caused by air layers under the surface of a leaf, which reflect light, or also by particular patterns in the hair on leaves, as in *Begonia* **species**. Variegation is often much sought-after by gardeners for aesthetic reasons because it gives a strong patterned look to the leaves, although it can be an indicator of pathology such as mosaic virus infection or a genetic mutation. It has been suggested that this may be adaptive, warning a possible predator that a leaf is apparently already under attack.

Weeds

Weed is a term for any plant that grows where it is not wanted. Even the most beautiful of plants can be considered a weed if it is growing somewhere that it shouldn't. While there is nothing objective that marks any particular plant as a weed, there are some plants that are more often viewed as such, either because they are very good at dispersing themselves or because they are very good at surviving.

Many common **garden** weeds are **annuals** or **ephemerals**, a strategy that depends on quickly producing vast quantities of **seed**. So, these kinds of plants easily perpetuate themselves during the growing season and additionally build up a seed bank in the **soil**, which then germinates year after year. Examples include chickweed (*Stellaria media*), meadow grass (*Poa annua*) and hairy bittercress (*Cardamine hirsuta*).

Other plants are commonly viewed as weeds because they are persistent. **Perennials** such as nettles (*Urtica dioica*), ground elder (*Aegopodium podagraria*) and hedge bindweed (*Calystegia sepium*) can regrow from the smallest section of **rhizome** left in the ground. They can even survive the **composting** process to be unwittingly spread around the garden.

Certain weeds are more than just an aesthetic nuisance. Japanese knotweed (*Reynoutria japonica*), for example, has such a strong **root** system that it can damage building foundations. The **sap** of giant hogweed (*Heracleum mantegazzianum*) can cause skin to blister badly in the sun. Both of these plants also happen to be examples of **invasive** non-natives; plants introduced from far-flung geographical locations, and which spread voraciously, untamed by any natural predators. These kinds of plants can cause huge ecological problems, outcompeting native plants, altering **habitats**, and endangering **ecosystems**.

History

Alexander von Humboldt

A German scientist and explorer, Alexander von Humboldt (1769–1859) pioneered studying the environment of the Earth as a whole. He mapped South America as he travelled, and described many **species** new to Western science. His central concern, however, was to describe how species related to the environment in which they flourished, and to unify scientific understanding under one heading, which was also the name of his overall publication: *Kosmos*. He published his discoveries with great passion and energy, and on large scale, and his huge fold-out diagrams of global temperature, magnetic measurements and mountain vegetation were instrumental in establishing our modern ideas about physical zones. Today we also see him as the first to explicitly identify human-made **climate change**, in the deforested parts of the Aragua valley, in Venezuela.

Ayurveda

Traditional Indian medicine begins with a system dating to the third century BCE known as Ayurveda, from the Sanskrit a*yus*, meaning "long life", and *veda*, meaning "knowledge". The focus on Ayurveda is on the balance of paired elements in the body, and any imbalance will give a greater tendency to a particular kind of illness. Such imbalances can be corrected by using Ayurvedic medicines, the huge majority of which are plant-based. The *Rig Veda*, dating to the later Bronze Age in first composition, names 67 plants and features 1,028 hymns about them, called *shlokas*, of which the most famous is the one called "Healing Plants". The most mentioned plant in the *Rig Veda* is *soma*, a sacred stem used in Vedic ceremonies but still unidentified; the second-most-mentioned is the peepul tree, under which the Buddha achieved enlightenment. Ayurvedic knowledge continued to develop over the years and the medieval pharmacopeia *Sarngadhara Samhita* prescribes medicines such as cannabis, camphor and opium.

Beth Chatto

Beth Chatto was an English horticulturist and **garden** designer who lived and worked in the twentieth and early twenty-first centuries. Chatto pioneered an approach to garden design which emphasizes that the best plants to grow are the ones that suit the conditions available. To reflect this, she coined the often-used maxim "right plant, right place". Her philosophy was shaped and is epitomized by the Beth Chatto Gardens near Colchester, the grounds of her family home – a site which Chatto developed over the course of 50 years. Rather than fight against the difficult conditions she found herself working with, Chatto drew on research into the wild origins of garden plants to choose **species** that were adapted to thrive in them, creating the likes of a gravel garden, a scree garden and a water garden.

Cacao

Theobroma cacao – from the Greek *theos*, meaning "god", and *broma*, "food" – is one of 17 species belonging to *Theobroma*, a **genus** of the mallow **family**. The cacao species itself comprises about a dozen significant **cultivars**. Cacao was probably originally domesticated for the soft pulp, rather than beans; the pulp can be eaten easily, whereas the beans must undergo a multistage preparation to produce cocoa. The ancient civilizations of Central America drank cacao as part of religious rituals and it was valuable enough to serve as a currency; it was first introduced to Europe after the Spanish were served it by Moctezuma. Like all significant modern food crops, the genetic diversity of cacao is a subject of research interest: wild crop genes could help to battle diseases caused both by monoculture of one cultivar and by **climate** crisis.

Carl Linnaeus

Carl Linnaeus was a Swedish scientist who lived and worked in the eighteenth century. His most enduring legacy is in the field of **taxonomy** and **naming**, where he formalized systems that are still in use to this day. Natural science in the early eighteenth century still largely relied on long and complex Latin descriptions to accurately identify species. Other scientists had attempted to simplify this system, but it was Linnaeus who honed what we now call binomial nomenclature and his *Systema Naturae* was the first publication to apply it consistently. *Systema Naturae* also introduced a hierarchical system of classification which, although markedly different to what we use today, was still an immensely influential idea. Over his lifetime Linnaeus used his new system to classify over 12,000 plants and animals, many of which retain the names he gave them to this day.

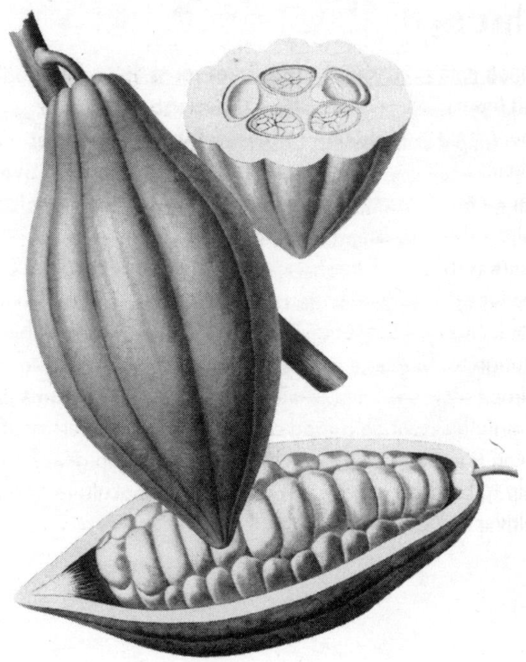

Cavendish banana

Cavendish is the name for a particular type of banana which accounts for almost half of all edible bananas grown worldwide. If you buy a banana in the supermarket, particularly in the UK and Europe, there is a very high chance that it will be a Cavendish.

Descended from a single plant grown at Chatsworth House – the ancestral home of the Cavendish family, from whom this banana gets its name – Cavendish banana plants give a good yield, producing sweet **fruit** without **seeds** which transport well. But these qualities come at a price. Lacking seeds, these plants cannot sexually reproduce. They must be propagated clonally, which means that plantations have a very low genetic diversity, leaving them vulnerable to **pests** and diseases.

The banana industry has already experienced the implications of this kind of homogeneity. In the first half of the twentieth century, most of the bananas we ate were of a type called Gros Michel. But in the 1950s, entire plantations were decimated by a fungal disease called Panama disease. The fungal **spores** of this pathogen persist in the **soil**, making it impossible to grow Gros Michel bananas in the same place without significant risk of infection. Cavendish became the dominant type purely because they seemed unaffected and could therefore be grown in infected soil. But a strain of Panama disease which does affect them has already been identified, and scientists worry that the Cavendish might go the same way as Gros Michel, with serious implications for the global banana industry and all the people who rely on it.

Charles Darwin

Charles Darwin (1809–82) was the English scientist who gave the most detailed first account of the workings of the theory of **evolution** in *On the Origin of Species*, published in 1859. Darwin was a keen student of botany and his works on plant fertilization and on orchids preceded the *Origin*, which itself contains much detail about how plants confirm the theory of evolution. He sought the advice and experience of gardeners, horticulturalists and botanists from around the globe, exploring subjects ranging from sea beans (**seeds** that germinate after floating in the ocean) and **carnivorous** plants to how **climbers** twine themselves around support. His **garden** and laboratory at Down House, south of London, are open to visitors, who can see the kind of experiments he undertook. His friendship with **Joseph Hooker** meant that they exchanged many letters, the originals of which were consulted in person at Kew by Benedict Cumberbatch when he played Hooker in the film *Creation* (2009).

Chinese herbal

Chinese herbal knowledge begins with the *Shennong Ben Cao Jing*, a compendium of oral herbal knowledge, said to have been written by the mythological Chinese ruler Shennong. Shennong allegedly invented technology ranging from the plough to the axe, as well as teaching humans to use plants medicinally, having tested them out on himself – one folk story says he once tested 70 different poisons in a single day. The *Shennong* includes medicaments derived from plants such as ginseng, rhubarb, ginger, liquorice and cannabis. A second mythological ruler, the Yellow Emperor, is the supposed author of another ancient Chinese wisdom text, the *Huangdi Neijing*, which is the central source for Chinese medicine. This was the first book to argue that diseases were caused not by demons, but by diet, lifestyle, environment and old age, and it is preventive in its tone and aims: "to administer medicines to diseases that have already developed is comparable to beginning to dig a well after you have become thirsty."

Cinnamon

Cinnamon is a valuable **spice** made from the **bark** of several different **species** from the **genus** *Cinnamomum*, one of the laurel **family**. The grade of cinnamon depends greatly on which species is grown, but its distinctive scent and taste always come from cinnamaldehyde, a yellow liquid that is contained in the inner bark of the **tree**. This compound is also present in other smells attractive to humans, such as almond and butterscotch. The flavour potential of cinnamon has made it highly valued throughout human history, and its origins were speculated on by the Greek Herodotus (fifth century BCE), who thought it was collected by giant cinnamon birds who made the sticks into their nests. In fact, the soft young stems are produced by **coppicing**, cutting the whole tree down to the ground every two years.

Coffee

Coffee is a stimulant drink made from the roasted and ground seeds of plants in the genus *Coffea*. The name stems from the Turkish word for the drink *kahveh,* and the **botanical name** for the genus comes from that. There are 124 species of *Coffea*, but only two are grown commercially in any great quantity: *Coffea arabica* and *Coffea canephora* (widely known by a synonym, *Coffea robusta*). Both are native to **tropical** Africa, but were translocated to commercial plantations throughout Southeast Asia, Central America and South America in the seventeenth and eighteenth centuries. It is likely that almost all the plants grown in the Americas are descendants of a handful of plants which came from Amsterdam's **botanic garden** in the early eighteenth century. This means that genetic diversity in plantations is very low, making plants susceptible to **pests**, diseases and the pressures of a changing **climate**.

Dinawari

During the medieval period much knowledge was generated and preserved by Islamic scholars within the Muslim empire. Three major figures – Dinawari, Ibn Juljul and Ibn al-Baytar – worked to collect and extend the understanding of plants, including how they grew from **seeds** or **roots**, and how to better care for them. Plants are mentioned specifically in the Qur'an, giving them special status as a symbol of paradise. Finding correct and universally applicable names for plants was an early preoccupation in the educated Muslim world, allowing for a useful standardization in medical and scientific writing and enabling a shift known as the Arab Agricultural Revolution. Probably the most famous figure in Islamic botany was Dinawari, and his book combined a herbal with poetry. There are many examples of Islamic texts building on existing knowledge, supplementing it with real-world observations from travellers and agriculturalists. The Muslim world added many hundreds of effective plant medicines to those known from the Greeks.

Dioscorides

Pedanius Dioscorides was a Greek physician who lived in the Roman Empire in the later first century, serving in the Roman army and compiling his masterwork, *De Materia Medica*, a work giving details of 600 plants that were seen as medically active. The book includes many plants still recognized as effective medicines today, such as opium, senna, colchicum and aloe. The book was the most important work in pharmacology for more than 1,500 years, and is the first acknowledged pharmacopoeia – a book giving information about medicines and how to make them. For many centuries it circulated in manuscript form, and information was often added from different medical traditions, such as Arabic or Indian sources. Later editions also included illustrations, to help identification, an important innovation. **Linnaeus** named the **genus** *Dioscorea* after him, the group including, appropriately enough, the very useful yam.

Domestication

Domestication is the process by which humans adapt wild **species** to our needs. It is the changes that we breed into organisms to make them more suitable for how we use them; the crops grown by farmers today are very different from their wild ancestors, selected over generations for characteristics which make them more productive and easier to farm. In plants at least, domestication begins with the bringing of a wild species into cultivation – the change from **foraging** a wild population to dictating where that population will grow. And in this sense, domestication is perhaps one of the most important developments in human history. It was a vital part of the process that transformed humanity from a nomadic hunter-gatherer society to a static agricultural one, laying the foundations for the modern world as we know it.

Dutch elm disease

A fungal disease, transmitted by tunnelling beetles, which kills elm trees. It was one of the first notable examples of internationally spread plant pathogens leading to the death and disappearance of a whole element of the wild landscape. The cause of the disease was identified in the 1920s by a group of seven Dutch scientists, all women. First seen in Britain in 1927, a new virulent strain was imported into England from Canada in the late 1960s, and within 10 years had killed most of England's elms. A small body of the **trees** survive in Brighton and East Sussex, where the council has taken a militant stance to maintain its disease-free status, and there are now more elms there than there were before the disease hit.

Forest bathing

Much recent scientific research has focused on the way in which humans gain positive advantage from their relationships with plants. "Forest bathing", or *shinrin-yoku*, is a Japanese idea, developed in the 1980s after government studies suggested there were positive mental and physical health benefits to be derived from contact with nature. There is some evidence that human beings improve in mood for significant periods of time after spending even a few hours in a natural setting. Nature Deficit Disorder has also been much discussed, especially with respect to children who no longer have as many opportunities to be outside. Even NASA has recognized the value of being around plants in creating a psychological sense of wellbeing to astronauts on long spells in space (see **Astrobotany**, page 184).

Francis Masson

Kew's first official plant explorer, Francis Masson (1741–1805) trained as a gardener and was picked by **Sir Joseph Banks** to sail with Captain Cook on HMS *Resolution*, landing in South Africa in 1772 (and not returning to England until 1775). During this time, the Scotsman collected and brought back many **species** new to European cultivation, such as *Kniphofia*, the much-loved **garden** plant red hot poker, and *Strelitzia reginae*, the bird of paradise plant, which he named after George III's wife, Charlotte of Mecklenburg-Strelitz. The *Encephalartos altensteinii*, a **cycad** he collected for Kew, is very slow-growing and remains in a container in the **Palm House**; it is thought to be one of the oldest pot plants in the world. His later expeditions didn't match the success of the first in Africa, and he died in Montreal just before Christmas 1805.

Garden

A garden is a place where natural elements, living or otherwise, are made to follow some conceived order. It may be either inside or more commonly outside, and will generally involve plants, although some Zen gravel gardens are for the most part stone. Gardens can be productive, and apparently completely utilitarian, but a brief look at any allotment will show that each vegetable gardener has a slightly different approach, and aesthetics almost always come into even the most functional space. Gardens are often ornamental and may include structures such as ponds, walls and arbours, to create height for growing **climbers** and to exclude the space from being overlooked. The idea of gardening seems to date beyond recorded history, to the very start of civilization: the biblical narrative in *Genesis* describes Adam and Eve leaving the Garden of Eden.

Greenhouse

A greenhouse is a structure in which plants can be grown so as to be protected from the elements. The idea of growing plants with protection is as old as the Roman Empire, when certain crops would be grown in covered containers – miniature greenhouses that could be wheeled inside to avoid the cold. But the concept has evolved significantly over the centuries, and the modern greenhouse is a dedicated structure that can be made from a variety of transparent materials (hence they are not always called *glass*houses) and can even have complex **climate** systems for creating the perfect conditions to suit whatever plants are being grown in them.

Among the most impressive greenhouses are those built in Victorian Britain. An influx of tender plants from around the British Empire coincided with advances in construction techniques and meant that such structures were built on an unprecedented scale, including great public glasshouses such as the **Palm House** and the **Temperate House** at Kew.

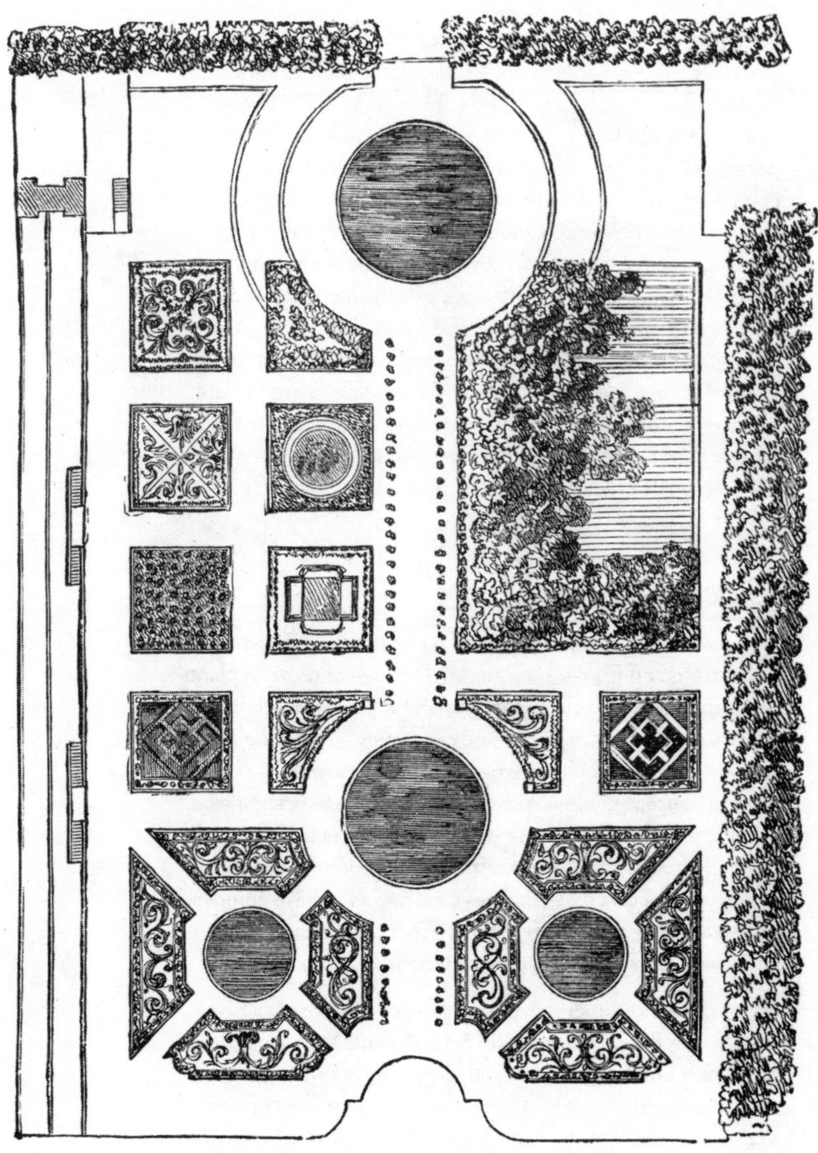

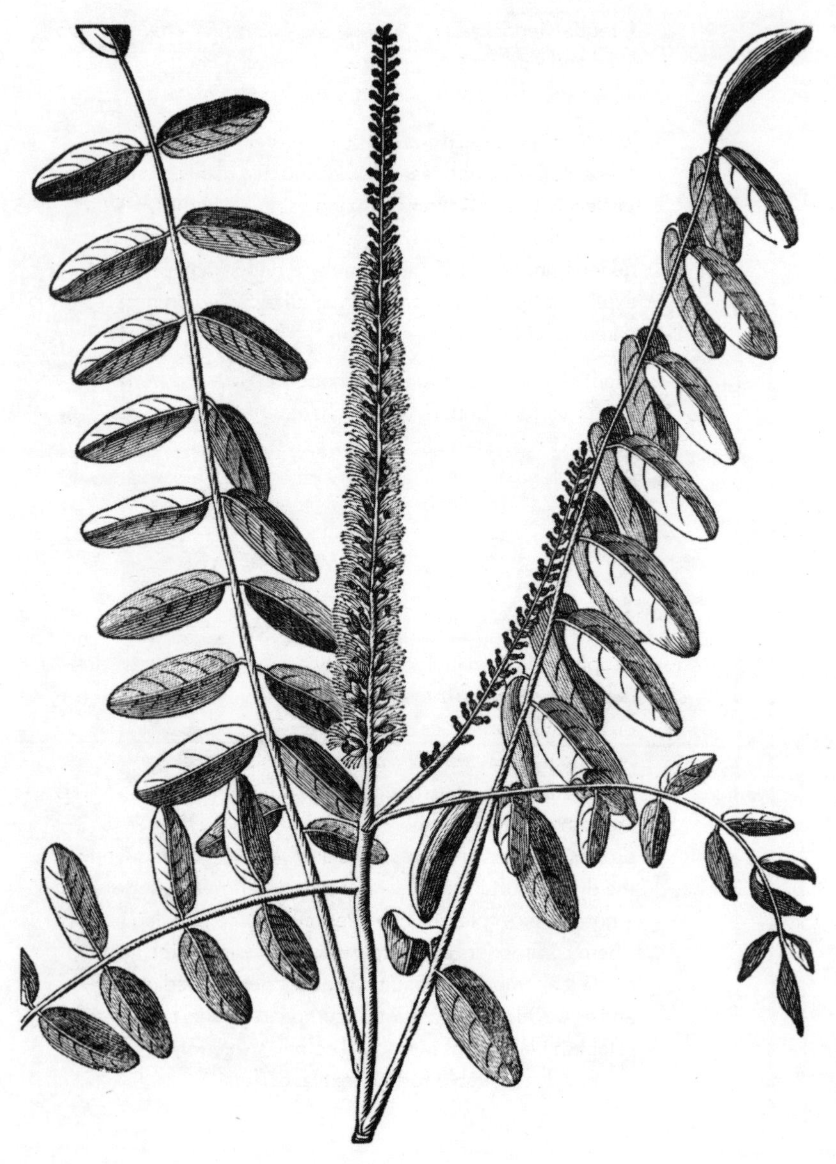

Gregor Mendel

Gregor Mendel was an Austrian botanist and clergyman who lived and worked in the nineteenth century. His theories on inheritance laid the foundations for modern genetics.

Working in the **gardens** of the monastery where he lived, Mendel spent eight years hybridizing pea plants, transferring **pollen** between **flowers** by hand using a paintbrush. With careful observation and meticulous record-keeping over generations of plants, he was able to deduce key laws that explained how characteristics might be passed on from parent to child.

Mendel's theories were such a huge step forward from contemporary thought about inheritance that they would be all but forgotten until 30 years after he published them. Although he hadn't called them genes, Mendel's work would later become the basis of this fundamental idea.

Indigo

People have used indigo for colour since ancient times, and there exist cuneiform instructions for dyeing with indigo on a Babylonian tablet from the seventh century BC. This blue plant dye is made from the **leaves** of plants from the *Indigofera* **genus**, most importantly *Indigofera tinctoria*. Blues are somewhat uncommon dyes in nature, which tends to produce shades of greens, reds and yellows. Consequently the dye was often strongly associated with wealth and status – notable examples include West Africa and also Japan, where commoners could not wear silk. Indigo's high value made it an important trading product during medieval times, and as world travel increased, European powers began to establish plantations in tropical colonies. Indigo was one of the few dyes suitable for the textile, cotton.

Joseph Banks

Joseph Banks (1743–1820) was an English landowner with great energy and enthusiasm for creating networks, collections and knowledge resources, during the reign of George III. As a young man Banks travelled with Captain Cook to Australia, and was instrumental in advising the king to colonize that continent. Banks was passionate about steering Kew to become a world-class scientific institution, swapping plants and **seeds** around the globe. He showed the interests and preoccupations of his time in researching cheaper food for enslaved people, to lower the cost of **sugar** production. Banks' own collections, held in central London as a reference resource, eventually became a founding part of the British Museum. He was the first European to write down observations of the animal he called "kanguru".

Joseph Hooker

Joseph Hooker was one of the greatest English botanists who lived and worked in the nineteenth century. His reputation was built on the expeditions that he undertook to Antarctica, the Himalayas, and the Middle East: journeys on which he recorded and collected numerous plant **species** new to science. A close friend and confidante of **Charles Darwin**, he also made important contributions to young scientific ideas such as **evolution** and **phytogeography**. Botany was a family pursuit: Joseph's father, William Hooker, was the first ever director of the Royal Botanic Gardens, Kew, a post that Joseph would later assume himself.

Among Hooker's most enduring horticultural legacies were the species of rhododendron that he recorded on an expedition to Sikkim in the Himalayas. Plants and seeds sent back to Kew fed a Victorian obsession with rhododendrons and formed the basis of a collection that can still be found in the **gardens** at Kew today.

Marianne North

Marianne North was an English artist who lived and worked in the nineteenth century. She is primarily known for the collection of her botanically themed oil paintings which resides in a dedicated gallery at the Royal Botanic Gardens, Kew.

The North family's wealth meant that Marianne was able to travel extensively. In her later life she undertook long journeys to places including Jamaica, Brazil, Japan, Borneo, India, Australia and South Africa – and this at a time when the steamboat was the quickest means of travel available.

North constantly painted what she saw. And much of her work is focused on the plants that grew in the places she visited. Although she did not adhere to the traditions of formal **botanical illustration**, the impressive detail recorded in her paintings and in her notes meant that they became important records of plants in their natural **habitat**. She was also the first to record and collect a number of **species** entirely new to science, including some that were described by botanists from her paintings, and which are named in her honour.

North donated her paintings to Kew and paid for the gallery in which they are housed. She intended them to educate and inform; in an age when travel was the preserve of the wealthy few, her works allowed the Victorian public to travel vicariously and thus to learn about the natural world.

Monkey puzzle tree

Monkey puzzle is the **common name** for a **species** of **tree** with the **botanical name** *Araucaria araucana*. A distinctive **evergreen conifer** with spiky, scale-like **leaves** covering its branches in their entirety, its name was coined in the nineteenth century when it was observed that even a monkey would have a hard time trying to climb it. The monkey puzzle's unusual appearance has led to its adoption as a popular ornamental **garden** plant, and it produces edible **seeds**. But it is particularly interesting because it has remained largely unchanged for millions of years. Fossilized remains of very similar plants have been found dating back to the Jurassic period. This means that trees very like the monkey puzzle have been growing on the Earth since the time of the dinosaurs.

Orchidelirium

Orchidelirium describes a period in the nineteenth century when the fashion for collecting orchids reached extraordinary heights. Many specific groups of plants attract especially obsessive collectors, and orchids are a good example. Collectors are particularly keen to have unusual new species, with rarity the crucial factor, and this peaked in Victorian times. As prices rose, plant nurseries and private individuals sent collectors to countries they knew were rich in wild plants, and much competition and skulduggery ensued. Novels and paintings of the time show that orchids were prized for their exotic appearance, frilled and decorated with patterns, a result of their coevolution with their pollinators. Today the orchid trade is strictly controlled by **CITES** and Kew is the first reporting authority for any suspected orchids seized at British ports.

Physic garden

A physic garden is a space in which **medicinal plants** are grown. A "physic" – probably from the old French word *fisique* – was a medieval term for a remedy or curative, and specific places set aside to grow plants for use in healing originated in the gardens of medieval monasteries. From the sixteenth century, as information about the medicinal properties of plants became more widely disseminated in **herbals**, these types of gardens outgrew the monastery walls and sprang up all over Europe, often attached to the medical schools of universities. In these specialist gardens, plants were grown not only for use in medicine, but also for studying and teaching. In this respect, physic gardens were the predecessors of modern **botanic gardens**.

Pteridomania

Pteridomania describes a particular enthusiasm for **ferns** which flourished in Victorian Britain. Coined by the author Charles Kingsley and originating in the Greek for "fern", *pteris*, the term encompassed documenting, collecting and cultivating ferns. The scale of interest in collecting live specimens was so intense that the populations of several species became endangered in the wild. In this sense, Pteridomania was a reflection of the Victorians' general passion for natural history, which often manifested as a desire to capture and catalogue rather than to conserve.

The influence of this fascination can also be seen in the formation of fern societies and the publication of fern books, as well as in the proliferation of the fern as a motif in art and design of the time, as evident on custard cream biscuits to this day.

Quinine

A plant **alkaloid**, quinine was likely originally known from traditional medicine in Peru, where the dried ground **bark** of the cinchona tree was used as a remedy for malaria. Spanish missionaries who had visited the new colonies in South America returned to Spain with cinchona bark, an important improvement on existing European treatments. The drug quinine was isolated from the bark in 1820, and remains in use today. **Linnaeus** named the bark in reference to the Countess of Chinchon, who was apparently treated successfully with it and introduced the bark to Europe — a story that was debunked long ago. However, quinine certainly is the first known use of a chemical compound to treat an infectious disease. It has strong side effects, but it remains a very important medical treatment, some four hundred years after its first introduction to Europe.

Rubber

Rubber is a stretchable waterproof product usually made from the latex collected from the Amazonian rubber tree, *Hevea brasiliensis*. *Hevea* is a spurge, a member of the euphorbia **family**, whose members almost all exude a milky viscose liquid when cut. These properties were first discovered by the Olmec, Maya and Aztec peoples of Mesoamerica.

Rubber tappers cut the bark to allow the latex to drain out into containers. The latex is then processed to make it ready for use. Joseph Priestley seems to have pioneered using "rubber" to erase pencil marks (hence the name), and it was Charles Goodyear who introduced vulcanization for tyres as a commercial process in 1839.

In Victorian times, South America was the main source for unrefined rubber, and Henry Wickham made it his unlawful mission to obtain **seeds** in order to grow rubber plants in the British Empire, rather than paying for it. He smuggled 70,000 rubber seeds from Brazil to Kew, which then sent out seedling plants to India, Sri Lanka and Malaya, to circumvent the South American monopoly. In modern times, rubber can be synthesized, but about half the rubber produced comes from living plant sources, mostly clones of high-yield varieties grown in plantations where rainfall and sunlight are sufficient. Rubber tapping is a skilled art: too shallow a cut will not harvest all the available latex, but too deep will harm or even kill the **tree**.

Rubber can be made from other latex-producing plants; when supplies were interrupted during the Second World War, both the Allied and Axis powers manufactured experimental rubber from the milky latex produced by cut dandelions. Recent experiments in Europe have enabled researchers to switch off the gene that makes the cut heal over, enabling more latex to be produced.

Saffron

The world's costliest **spice**, saffron comes from the brightly coloured stigma and style of the saffron crocus; it has been cultivated for so long that the plant's wild origin is no longer known. The high market price results from the number of **flowers** used – about a hundred and fifty thousand blooms per kilo of finished spice – and the delicate process of harvesting the **stamens**. The original "threads" have a very vivid red colour from the red stigma, but saffron tends to colour food golden yellow, via a carotenoid plant pigment called crocin; the delicate flavour comes mainly from a compound called safranal, and is greatly enhanced by the harvest being carefully dried. Saffron is graded when it comes to market, and the grades relate to how much of the stigma and style relatively are included in the end product.

Silk Routes

The Silk Routes were an ancient network of trade routes linking Asia and Europe. While the eponymous trade in silk was undoubtedly lucrative, it does not represent the full picture; in the centuries surrounding the beginning of the common era a huge variety of commodities were traded between East and West, a pattern of exchange that was hugely influential in shaping many aspects of global human society.

Many of these traded goods were plant products – spices and cotton, wine, and incense. Food crops also began to travel; pistachios, walnuts and almonds first came to Europe along the Silk Routes. Perhaps most surprisingly, the apple – a **fruit** that is perhaps considered typically English or American – has been shown to originate in wild populations in Central Asia, and looks and tastes as it does today thanks to its translocation on these ancient trade routes.

Spice

Spices are parts of plants, generally dried, used to flavour food. They are usually **seeds**, berries, **bark** and **roots**, and should be seen in contrast to herbs, which are the **leaves**, stem and **flowers**. Historically, a major source of spices was the Indian subcontinent, and it remains the supplier of a majority of spices worldwide. The spice trade drove much early exploration and travel, as the remarkable prices commanded for such products made shortcuts and alternatives worth seeking. World cuisines often make distinctive spice mixes that can instantly be recognized as typical of a particular country, ranging from jerk spice, comprising Scotch bonnets and allspice as a base, to chaat masala, with its particular tang that includes familiar Indian flavours such as coriander, as well as amchoor (dried mango powder) and asafoetida, a dried latex with a pungent smell.

Sugar

Sugar is a product that can come from several different sources in the plant world. Sugar cane is a tall member of the **grass** family, grown in **tropical** conditions where there are many hours of sunlight a day and sufficient rainfall; the other main agricultural source of sugar is beet, which can be grown in much higher latitudes and heavier **soils**. Sugar cane is the world's most grown crop measured in tonnes, with a total of 1.9 billion tonnes produced in 2020. It is used directly to eat but also to produce alcohol and as a biofuel, especially in Brazil. The dominance of sugar as a crop came about as European diets became much sweeter during the eighteenth century, but the labour-intensive crop was produced by the work of enslaved people.

Tea

Originating from China, tea is the drink made from the new **leaves** of the **evergreen** shrub *Camellia sinensis*. Tea is consumed by two-thirds of the world's population for its mellow taste, but arguably more importantly for the caffeine the leaves contain. The plants that give tea its taste are all originally one **species**, but have diverged into strongly identifiable varieties with distinct flavours. Chinese Assam and Indian Assam are estimated by genetic studies to have last had a common ancestor around 3,000 years ago, and were a luxury good traded across long distances since at least that time.

As European society became more well-off and commercially orientated during the seventeenth century, tea became a very popular drink, especially in England. Commercial imperatives led the British to pursue the idea of possessing Chinese tea plants, for the cost of imports created a trade deficit with China, which had already led to the Opium Wars. As a result, in the 1830s the British began trying to grow tea in India, leading to a variety of underhand efforts to spirit tea plants out of China. Eventually, the British settled on a mix of Chinese leaves and the local Assam tea of northern India. The blend of teas, and the ways in which the leaves are prepared, can make the flavour vary considerably, and human tastes in tea have always been very particular: as early as the second century BCE, the Chinese general Liu Kun wrote home plaintively asking to be sent some "real tea".

The Palm House

The Palm House is a glasshouse situated at the Royal Botanic Gardens, Kew. It was built in the 1840s to display Kew's growing collection of **tropical** plants. As the British Empire expanded, collectors were sent out to retrieve plants and **seeds** from its far corners. The specimens that were sent back from the tropics could not be grown outside in the UK, so specialist environments had to be built to house them. The conditions inside the Palm House mimic those of a **rainforest**. This means that it is kept warm and humid. Conditions are perfect for many well-known plants including **bananas**, **coffee**, pepper, sugar cane and cocoa, all of which can be found growing in the Palm House.

The Temperate House

The Temperate House is a glasshouse situated at the Royal Botanic Gardens, Kew. It is so named because it houses plants that grow in **temperate** regions – those areas of the world with moderate **climates** – but which cannot be grown outside in the UK because they do not like colder temperatures. The Temperate House contains more than 1,500 different plant species from five of the world's continents, including numerous plants that are extinct in the wild.

The building was originally finished in 1899 after almost 40 years of construction, and by the early twenty-first century its fabric was starting to fall apart. In 2012 work began to fully restore the entire building. Almost every plant (with the exception of seven **trees** too large to move) was propagated or removed from the building before being replanted once the work was finished six years later.

Theophrastus

Theophrastus, a nickname meaning "phrased in a godly way" to describe his way of speaking, studied with Aristotle, and eventually became his teacher's successor as *scholarch* or head of the philosophy school in Athens. It was Aristotle who gave him his nickname. Few works of the hundreds thought to have been by Theophrastus survive, but two on botany – *Historia Plantarum* or *Enquiry on Plants*, and *On the Causes of Plants* – were important for a thousand years after his death, and **Linnaeus** called Theophrastus the "father of botany". *Historia Plantarum* details the different categories of plants and how to grow them, and is mostly about economically important **species**; Theophrastus collected information from those who had travelled to Asia with Alexander the Great, and gave the first European reports of pepper, **cinnamon**, frankincense and myrrh.

Tobacco

Introduced to Europe from the Americas after the first contact between the cultures, tobacco quickly became an important element in transatlantic trade. Cultivation was intensified by the slave trade, and racial justice remains an issue in tobacco farming. It is made from the **leaves** of the tobacco plant, and can come from several species in the **genus** *Nicotiana* of the nightshade **family**. Up to seventy species are smokeable, and there are other differences depending on the way in which the leaves are grown and prepared – for example, shade-grown tobacco is kept under tenting to reduce direct sunlight, which changes the flavour and colour of the leaves. The nicotine in tobacco is the addictive compound, but the leaves can be cured in different ways to change the balance between sugars and nicotine content.

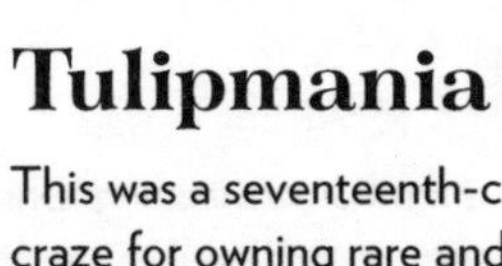

Tulipmania

This was a seventeenth-century Dutch craze for owning rare and exclusive varieties of tulips, in the form of **bulbs**. It began with enthusiasm for the tulips themselves, but quickly became an arena of financial speculation as bidders and sellers vied to profit from the rapidly rising prices. Tulips were introduced to Europe from Iran, probably via the Ottoman Empire, in the mid-sixteenth century, and became a luxury item in the Netherlands, which was a rich, world-leading economy and was quick to embrace consumer culture in the form of **garden** plants. The particularly desirable **flowers** had bizarre and unique patterns created by mosaic viruses, a group of diseases that cause complex mottling to develop. Parrot tulips today are affected by the same tulip breaking virus (TBV), but modern **propagation** techniques now allow their bulbs to be produced at normal market prices.

Vasculum

The vasculum is a piece of specialist equipment popularized by nineteenth-century plant-collectors which allowed them to store freshly gathered plants in an over-the-shoulder container during a day in the field. Its name comes from the Latin for "small vessel". Vascula were usually made of metal, either lacquered or painted, to keep the specimens cool and protected while they were transported back to base, and were available in sizes ranging from very portable to those able to carry a large plant. **Joseph Hooker**, later Director of Kew, took two on HMS *Erebus* as detailed in his Antarctic Journal, along with a pair of **Wardian cases**. The vasculum has now fallen out of use and botanists today are likely to prefer a combination of tight-lidded plastic containers and newspaper, with Tupperware a firm field favourite.

Wardian case

A Wardian case was a container made of glass and timber, used in the nineteenth century to house live plants. It was invented by Nathaniel Ward, a doctor and amateur naturalist, and used in domestic settings to grow more sensitive plants, such as the **ferns** that were so popular in Britain at the time. But it became particularly important for the movement of plants which it enabled. In 1833 one was used to successfully transport a selection of live plants from London to Sydney – at that time, a journey of eight months – and to likewise return to England with a selection of Australian **flora**. Over the next 100 years it was used to translocate a number of important crops, which had huge economic and societal impacts, both good and bad. It also made possible the creation of some of the great botanical collections we enjoy at places like Kew today.

Wollemi pine

In the wild, the Wollemi pine (*Wollemia nobilis*) is probably one of the rarest **trees** in the world. The **genus** was thought extinct, known only through fossils. But in 1994 a small group was found growing in a remote part of Wollemi National Park in Australia, from where the **species** got its name. With less than 100 individuals surviving in the original population, the species remains critically endangered in the wild. But in cultivation, **conservation** efforts have met with enormous success. A new population has been planted, specimens have been distributed to **botanic gardens**, and Wollemi pines are now sold commercially, distributing the species in gardens worldwide. Perhaps the Wollemi pine's most striking feature is its mature **bark**, which is deep, dark brown and has a texture like bubbles.

Documentation

Arboretum

An arboretum is an area dedicated to the cultivation and display of **trees** and shrubs. In Latin, *arboretum* means "a place planted with trees". But it was John Loudon – an influential nineteenth-century botanist and writer – who first used the term in English to describe a curated collection of trees, like the one he planted at Derby Arboretum, Britain's first public park.

Like **botanic gardens**, arboreta are defined by their purpose – containing a breadth of **species**, growing rare or unique trees, and recording and labelling the specimens that are there. They are intended to enable scientific study, to educate, or to aid in **conservation** efforts. They are often found as part of larger botanical institutions, where they are distinguished within the wider collections by their focus on woody plants.

Botanic garden

A botanic garden is a place where collections of living plants are grown and displayed, but they are differentiated from other gardens by their purpose; the plants are grown for more than just aesthetics. This purpose can vary – it might be scientific research, education, or **conservation** for example – but the common theme is always a focus on the plants themselves. This is reflected in the most fundamental feature of all botanic gardens; that the plants are accurately identified and labelled with their scientific names. They are usually linked to a catalogue of records which give detailed information on that plant: where it has come from, the environment in which it was found if it was collected from the wild, how it has been grown since it arrived in the garden.

Botanic gardens grew out of the **physic gardens** of the sixteenth century, which were resources for the teaching of medicine. As plants began to be transported around the world in the age of empire these gardens expanded, taking in ever more diverse plants with less emphasis on their **medicinal** properties and more on their economic value. Through the enlightenment – as botany became an endeavour in its own right – botanic gardens came to be centres for the advance of plant science, a role which continues to this day. Increasingly they also play an important role in plant conservation, not only providing safe havens where endangered **species** can be grown, but also providing the space for our knowledge of the plant kingdom to be built so that we have the means to protect **biodiversity** and tackle global issues such as **food security** and **climate change**.

Botanical illustration

Botanical illustration is the art and science of producing images of plants by hand, to help identify specimens. It uses a particular specialist set of skills, and the end goal is not a beautiful image – although many such images are aesthetically pleasing – but instead to allow botanists to gain as much information about the way the plant grows in the wild as possible. It may seem surprising that artistic illustration continued after the advent of photography, but artists can make choices about how to show the plant in detail, which are not possible when photographing a life form in real time. A great botanical artist can capture the habit and character of a plant, and the end result can show either many different stages of a plant's life cycle at once, or the detail of a single, especially significant stage. Normally a **species** will be identified by a **type specimen** that is a dried plant, but some plants are so delicate that this does not work, and the type specimen is then likely to be a botanical illustration.

The time and knowledge involved in producing such images makes them valuable, although in earlier centuries botanic artists were often highly skilled artisans who rarely had the opportunity to see the plants in the wild. Many of the profession's most notable practitioners were women, including Maria Sibylla Merian (1647–1717), who began to publish her illustrations when she was in her mid-twenties. In her early fifties she travelled across the globe, making "a long-dreamed-of journey to Suriname" to study and paint **tropical** insects in their natural setting.

Botanical name

A botanical name is the formal scientific name for a plant species. Botanical names are particularly useful because they are unique, standardized and universal. Applied correctly, a botanical name refers to the same kind of plant anywhere in the world and will be the only botanical name by which that plant is known.

The botanical name for a plant will have at least two parts. For example, the botanical name for the common daisy is *Bellis perennis*. *Bellis* tells us the **genus**, and *perennis* the species. They are italicized to separate them from surrounding text, and the genus name is given a capital letter. A formal botanical name should also include the "authority" which is an abbreviation of the name of the scientist responsible for naming and publishing the species e.g. *Digitalis purpurea* L., where L stands for **Carl Linnaeus** (see **Naming**, page 139).

Botanical names really only apply to plants that can be found naturally in the wild – plants that are produced by commercial plant breeders are named in a different way (see **Cultivar**, page 45).

CITES

CITES stands for the Convention on International Trade in Endangered Species of Wild Fauna and Flora. It is an agreement between the governments of 184 countries and aims to protect species from **extinction** by regulating trade so that it does not damage wild populations. High demand for a plant or animal product can result in wild populations being diminished quicker than they can recover. CITES regulations restrict trade in these products to ensure trade is legal, traceable and sustainable.

There are three different levels of regulation (called Appendices) protecting more than 30,000 species of plants. Perhaps the best known are timbers, particularly of the genera *Diospyros* (ebony) and *Dalbergia* (rosewood) which are often used to make furniture and musical instruments. Many horticultural plants are also regulated by CITES, including orchids, **cacti**, snowdrops and cyclamen.

Collecting plants

The majority of plants grown in European gardens today were brought from elsewhere by collectors. This is especially true in the United Kingdom, which has a particularly favourable mild and maritime **climate** where introductions flourish. Collecting in the past was often done by adventurers, paid to collect, although many significant plant discoveries were made incidentally to their work by missionaries, sailors or soldiers. Networks of local intelligence were extremely important, and collectors with native language skills who enlisted help from local peoples with knowledge of the landscape and wild **flora** were invaluable. However, these resourceful individuals were rarely reflected in the naming process, which tended to favour elite figures from the British Empire's centre.

Plant collecting reached its height of fashionability in the eighteenth and nineteenth centuries, when thousands of plants were brought to Europe from around the world and introduced into garden cultivation. At the same time naturalists began to form collections, in which locating and describing **species** new to western science was an important goal of research. Effort went into creating systems that could order and regulate such collections, from techniques used to keep a plant sample fresh (see **Wardian case**, page 122) to the rules that dictate its naming (see **Naming**, page 139).

Today, there is a much stronger sense of the ethical obligations in collecting plants that may eventually bring great profit, such as in the form of lucrative medical drug research. Plant collecting is now boundaried by strict contractual agreements drawn up before collectors set off. These are determined in shape by the Convention on Biological Diversity, which aims for equitable sharing of any economic benefits arising from any biological materials collected. Nonetheless poaching continues to be a problem, especially in plants that attract obsessive collectors. In 2014, a thief stole from Kew one of the last hundred plants left globally of the tiny waterlily, *Nymphaea thermarum*.

Common name

A common name is any name for a plant which is not its formal **botanical name**. For example, "daisy" is a common name for the plant with the botanical name *Bellis perennis*. Common names are useful because they are usually widely understood and easy to remember. But they can also cause confusion. There is no system governing them, and they are often in the vernacular, so the same plant might have different common names in different places, or two different plants might have the same common name. The most often cited example is the bluebell. In England this is widely understood to be the much-loved spring **woodland** plant with the botanical name *Hyacinthoides non-scripta*. But in Scotland a bluebell is a different plant altogether: *Campanula rotundifolia*, which is found in **grassland** and heath in late summer.

Cyanotype

Cyanotype, from the Greek *kuáneos*, "dark blue", and *typos*, a "mark" or "impression", is the word for a distinctive, simple method of making an image of plants by photographic means. When it was new and fashionable in early Victorian times, botanists keenly embraced the detail and subtlety it offered, leading to a style that is unmistakable. Anna Atkins, the Victorian botanist and photographer, was particularly skilled at making cyanotypes, especially of seaweed. She is arguably the first person to publish a book containing photographic images, in her 1843 *Photographs of British Algae: Cyanotype Impressions*. The book was produced for private use, and only 17 copies are now known to exist: as a result, in 2004 a set sold at auction for £229,250.

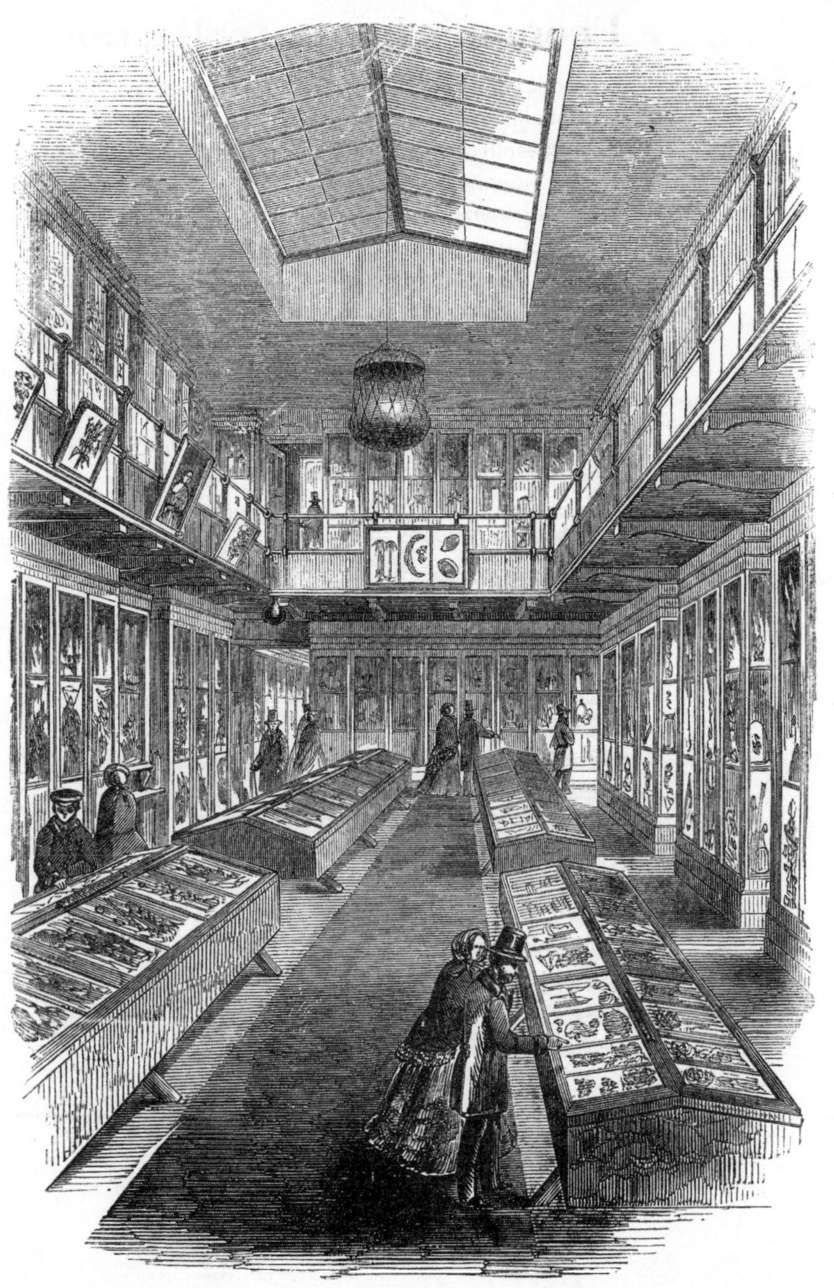

Economic botany collection

Economic botany is the study of the useful advantages of plants. Interest in economic botany flourished during the eighteenth century, when government administrators became increasingly aware of the range of benefits that might arise from gathering local knowledge about the useful properties of plants. In particular, economic botanists sought examples of plant materials in use, collecting items ranging from walking sticks to sieves to poisons. The Victorians were especially keen on this kind of accumulation of knowledge, and at Kew a separate collection was established in 1847 for just such a purpose, with a smart museum to display its assets. This museum, though, was not just a storehouse but the centre of a system of circulation, as items were sent and exchanged with many other **botanic gardens** and museums, with the aim of increasing knowledge, and bringing tangible benefits.

Family

In the ranked hierarchy of **taxonomy**, family is the rank above **genus** and **species**. A family can be thought of as a grouping of genera with a set of shared characteristics. Families can therefore be made up of quite a wide variety of different plants. But many families are still recognizable by distinctive features, particularly in **flower** form. Families are referred to by names that end with "aceae", which comes from the Latin meaning "resembling". For instance, Asteraceae or the daisy family. There are currently 452 accepted **vascular** plant families, although this number is constantly being debated and revised.

Flora

After the Roman goddess of plants, flora is a word used to refer collectively to the plant life of an area or period of geological time – for example, flora of the British Isles, or flora of the Jurassic period. It means the naturally occurring or native plant life and, although it is a collective word, in scientific usage it implies specificity: flora means the particular **species**, not the kinds of plants. For example, while it can be said that British flora includes **tree** species like *Quercus robur* (English oak) and *Quercus petraea* (sessile oak), it would not make sense to say that it includes oak **woodland**. Flora is also used as a title, capitalized, for a particular kind of publication which catalogues these plants, a published record of that community of plant species.

Florilegium

A florilegium in botanic terms is a very large-scale book with colour images, designed to illustrate ornamental plants, as opposed to medical and **herbal** texts devoted to useful plants. The skills of botanic artists to prepare scientific illustrations were harnessed in florilegia, to produce very costly and high-status volumes. Even when the illustrations were printed, they would have been hand-coloured, a process that requires great skill and painstaking patience. The term can refer to illustrated manuscripts about plants from medieval times onwards. Today florilegia are rarely produced, requiring wealthy patrons, and a notable recent example was the one made for Prince Charles's garden at Highgrove in 2008 and published at elephant folio size, meaning that special supports are required to avoid damaging the binding when opening the book.

Fungarium

This word refers to a reference collection of **fungi**, and is the fungal kingdom's equivalent to a **herbarium**. A collection of this kind aims to house **type specimens**, which act as examples for any scientist aiming to define or identify an individual species of fungi. In the modern era, such collections also take into account DNA evidence, which can enable much more precise assessments of the closeness of evolutionary relationships, sometimes radically reshaping the existing family tree. A fungal collection mainly comprises the dried fruiting bodies of fungi, but it can also include fungi that are microscopic in size, such as plant pathogens like downy mildew.

Genus

A borrowed Latin word, genus is the rank above **species** in the hierarchy of **taxonomy**. Individual plants in a genus have specific and unique characteristics in common to be identified as a group but, unlike those of a species, are not necessarily capable of interbreeding.

A plant's genus is used in conjunction with its species for a **botanical name** (i.e. *Geranium macrorrhiza*). All the different species of *Geranium* make up the genus *Geranium*. So if we aren't sure of the species, a genus name can be used by itself to narrow it down: "It's some sort of *Geranium*, but I'm not sure which one".

Herbal

A herbal is a written record of plants and their properties, particularly focused on **medicinal** usage. Herbals contain descriptions and illustrations to help identify useful plants, as well as instructions on how they might be used. Many ancient healing systems were based in herbalism, and this knowledge has been recorded for thousands of years in documents that could be described as herbals. Manuals of useful plants and alternative medical remedies exist to this day. But the great age of the herbal was Europe in the fifteenth century, following the invention of the printing press. Among the first documents to be printed, these volumes were largely reworkings or compilations of earlier manuscripts but were important reference books for physicians of the time. They formed the foundations of modern botany, particularly in their classification of plants.

Herbarium

A herbarium is a collection of plant material which provides a physical reference for the study and identification of species. A herbarium is organized in a storage building according to a classification system, so that closely related plants are near to one another and can be compared. Each species has a holotype specimen, originally used to define the species when first described and named and which can be used as a point of reference for any disputed identifications. Today, DNA evidence is increasingly important in elucidating evolutionary relationships between different groups of plants and shedding light on complex parts of the plant family tree. Older records are still important, though, and are now being used in novel ways, such as to track changes in the geographical distribution for species, providing evidence about **climate change**.

Kew's Herbarium is one of the largest in the world and contains about 7 million specimens, of which 330,000 are **type specimens**. It grew from an original private collection that belonged to **Sir Joseph Banks**, and was made official at Kew from 1841. Kew now adds about 25,000 new specimens a year, which are collected by Kew staff but also by its scientific partners, in a worldwide network of exchange. This allows botanical institutions globally to build collections, compare and contrast specimens, and fill gaps. The collection is tightly atmospherically controlled to protect it, and new additions are unwrapped in a cold storage area to reduce the risk of introducing insect or other plant **pests** and pathogens, which are a threat even to dead plant tissue.

Living collection

A living collection is a gathering together of living organisms. Individual living things, which can be conceptually considered as a group, cared for by one institution. Just as a collection of historical objects is held, curated and displayed by a museum, living collections also exist in specialist institutions. **Botanic gardens** and zoos are stewards of different kinds of living collections. Living collections rarely exists merely for their own sake. They are usually resources for research, for **conservation**, for education. The plants in a private garden are not generally considered a living collection, but all of the plants at Kew – some 27,000 different taxa, which are used regularly by scientists and educators – make up its living collections, the most diverse collection of plants in the world.

Naming

When taxonomists decide that a plant is different enough to constitute a new **species**, it must be given a formal **botanical name**. But for botanical names to be useful, this naming process must be governed in some way. So, there is a code which dictates what taxa can and should (officially) be called, and conventions on how these names are conferred.

These conventions are set out by the IAPT (International Association for Plant Taxonomy) in its code, the International Code of Nomenclature for **algae**, **fungi** and plants. This outlines the process that must be followed for a name to be accepted, and it also sets the rules which the name itself must abide by; it must be binomial (that is, made up of a **genus** and a species name), it must be unique and it must be grammatically correct (in Latin). Inappropriate or offensive names are not allowed, and you cannot name a new species after yourself.

Apart from these few stipulations, there is relatively free choice. But the whole idea of naming is simplification, giving us easier ways to refer to things. So, names are often chosen with a view to saying something about a species, thus helping recall. For instance, this might be where it comes from (*Chaenomeles japonica* comes from Japan), or it might be the colour of its **flowers** (*Alcea rosea* has rose-pink flowers).

Despite all the rules, scientists still manage to have fun with the names they confer. A new species of begonia with nearly black **leaves** and bright red flowers was recorded in Borneo in 2014. It was named *Begonia darthvaderiana* after the popular *Star Wars* villain who shares its colour scheme.

National Plant Collection

A National Plant Collection is a themed collection of plants registered with the charity Plant Heritage. There are over 690 registered collections, each of which represents a different group of plants. It might be plants of a particular **genus**, like different **species** of tulip. Or it might be **cultivars** of a plant bred at a certain time in a certain place; *Dianthus* cultivars bred in the UK before 1970, for instance. Or it might be a collection linked by a specific person, such as **medicinal plants** used by Erasmus Darwin. Whatever the theme that ties a collection together, all the collections contribute to one aim: to conserve the diversity of garden plants. National collections are held in a variety of settings, ranging from public parks to commercial nurseries and private gardens, but all are available to view – at least, by appointment.

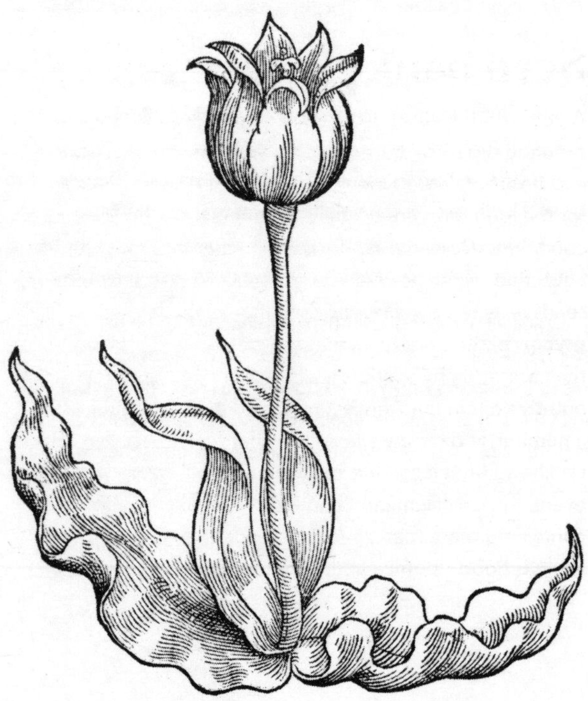

Red List

The Red List of Threatened Species is a database of information maintained by the International Union for the Conservation of Nature (IUCN). It records the risk of **extinction** faced by animal, **fungi** and plant species. Expert assessors look at various aspects of data on a species before assigning it a category. In ascending order of risk the categories are: Least Concern (LC), Near Threatened (NT), Vulnerable (VU), Endangered (EN), Critically Endangered (CR), Extinct in the Wild (EW) and Extinct (EX). This kind of information is essential to encourage, inform and direct conservation work that will protect species. In turn this helps to preserve **biodiversity**, which is increasingly recognized as fundamentally important to the natural world on which we all rely.

Seed bank

A collection devised in response to the idea of future-proofing botanical collections against **climate change** and **habitat** degradation, by storing quantities of **seeds** for the long term, in mostly low-temperature, low-humidity conditions. Research has focused on the viability of such collections when selected for **germination**, and different seeds require very different conditions, whether it is special **mycorrhizal** fungi to increase successful germination, as in the case of orchids, or particular actions to preserve the rich nutritive oils in nuts and acorns. Seed banks should exist in a number of different places, in order to increase the likelihood of at least one surviving a global apocalyptic event. Kew's Millennium Seed Bank at Wakehurst in Sussex, containing more than 2.4 billion seeds in an underground vault, is flood-, bomb- and radiation-proof.

Species

Species is the lowest rank in the hierarchy of **taxonomy**. It is difficult to objectively define exactly what constitutes a species. The most often cited definition is the so-called biological species concept, which states that a species is made up of individuals which are able to interbreed to produce fertile offspring. But even **Charles Darwin**, in his book *On the Origin of Species*, said "No one definition of species has as yet satisfied all naturalists" and there are currently at least 25 different accepted definitions of what the word species actually means.

Practically speaking, individual plants in the same species have a high number of common characteristics, as well as sharing the same requirements for growing conditions and other resources needed to thrive. Short of a clone, you cannot be more sure of what to expect. Scientists have calculated that there are more than 350,000 unique **vascular** plant species currently known to science, with many more being described at a rate of about 2,000 each year.

Type specimen

A type is a specimen nominated to represent what is meant by a particular scientific name. For a scientific name to be valid, it must have several things associated with it – a detailed description of the **species** for one. But since the 1930s, the International Code of Nomenclature (see **Naming**, page 139) has also called for a type specimen. Life is so varied, and there are so many different factors which can subtly affect how an organism looks, it is essential to have a visual reference. Something which we can point to and say: "that is what that particular scientist meant by that name". The most important kind of type specimen therefore is the holotype, which is the single specimen nominated by the scientist who first described the species and published the name. For plant species, types take the form of a dried specimen kept in a **herbarium**.

ENVIRONMENT
& ECOLOGY

Algae

Algae is a catch-all term used to refer to a wide grouping of organisms that **photosynthesize**. There is no one common ancestor of the algae, but there are some commonly recurring characteristics; the great majority of algae are **aquatic**, all of them photosynthesise. But algae lack true specialized structures that plants have developed, like **roots**, stems or **flowers**.

Algae is actually the plural form of the word, and the singular, *alga*, comes from the Latin for "seaweed". Seaweed is perhaps the most well-known of the algae, but there is a huge diversity of forms from single-celled organisms to seaweeds like the giant kelp, which can grow to more than 30m (100ft).

Atmosphere

Surrounding the Earth is a layer of gases and aerosols (tiny particles suspended in the air; substances such as water vapour, **pollen**, and ash), and this is known as the atmosphere. From the Greek words *atmós* (vapour) and *sfaîra* (ball), it is vital for life, holding the oxygen we breathe and the carbon dioxide that photosynthesizing organisms require, as well as blocking enough of the sun's radiation to make the surface of the planet habitable.

While the role of plants in maintaining oxygen levels in the atmosphere is overstated, what is true is that they take in and store a significant amount of carbon dioxide. The world's remaining **tropical forest** alone is thought to absorb up to 15% of our annual carbon dioxide emissions, helping to reduce our impact on the **climate** and making efforts to rein in deforestation all the more urgent.

Biodiversity

Biodiversity – a shortening of biological diversity – is a term which encompasses the enormous variability of life. From differences between individuals within a **species**, to differences between species, diversity is seen at every level of the living world.

High biodiversity is recognized both as an important indicator of **ecosystem** health, but also as a protection against ecosystem decline. Imagine a fictional ecosystem that is home to only one species of plant. As the single primary producer (see **Autotroph**, page 184), it supports all other life. The entire ecosystem is therefore dependent on that single species being able to adapt to any pressures. If it can't, the ecosystem collapses and all life dies. In an ecosystem that is home to a broad diversity of plant species however, variety gives more opportunities for adaptation and thus makes the system more resilient.

The natural world provides for humanity. Not only material things like food, textiles and medicines, but also indirect services like coastline and **soil** protection, water source filtration, even the oxygen that we breathe. But all of these things depend on the natural functioning of the Earth's ecosystems, complex and fragile things which are strained by numerous pressures including **climate change**, **habitat** loss, and pollution. Biodiversity builds resilience into ecosystems. The more complex and varied an ecosystem is, the better it can weather these pressures. Protecting biodiversity is therefore a way to protect these natural services that we rely upon.

But there is also an emotional element to this too. The variety of life is the aspect of the natural world which has most potential to delight and fascinate. How bare the world would feel if our exposure to nature was limited to just wheat fields and city parks.

Climate

Climate is the word that describes the meteorology or **weather** of a region over the longer term. Rainfall, average temperatures, **humidity**, level of cloud cover, winds and air pressure are all part of the picture, and knowing these factors gives scientists a good sense of what kind of plants are likely to be able to thrive in an area. Climate is also affected by a place's continental position, whether central or on a coastal edge, by the kind of landscape there is locally, by altitude, and by a place's position on the globe. Palaeoclimatologists use fossil and other preserved evidence to draw conclusions about the Earth's climate in the past; this can give a strong sense of how the climate has changed. Climate tends to be talked about as a 30-year average, but conditions are now shifting so fast that this may become outdated.

Climate change

Humans have affected the natural world in remarkable and significant ways, but none seems more desperately challenging than the concentration of greenhouse gases which is rapidly raising the average temperatures on our planet. There are several issues that are particularly significant for plants; they take carbon dioxide out of the **atmosphere** and store it; as a result, plants are the single biggest carbon sink on Earth. However, heat and water shortage both put stress on plants. Lack of water forces plants to shut their **stomata** and slow the process of growth, making it more difficult for them to sequester carbon via **photosynthesis**. Heat also changes the level of productivity at which plants can function. Plants also find it much more difficult to shift **habitat** in response to stress than animals do. These changes have led scientists to predict greatly raised rates of **desertification** and **extinction**.

Kew's work on climate change has focused for example on finding alternative genetic capabilities in food crops (see **Crop wild relatives**, page 73). Another strategy has been **seed banking**, done in order to preserve natural resources in case they become extinct in their wild habitat. Scientists worldwide must also work to assess extinction threats, so that **conservation** priorities can be identified. And science can also model how climate change might have an impact upon global **ecosystems**: for example, Kew has a monitoring programme in the Páramo in Boyacá, Colombia, to study this fragile and unique mountain habitat. Such a project has value in its own right, but also provides an example of good practice for other such studies in future.

Climatic zones

The Earth is divided by scientists into **climate** zones, for the purposes of making generalizations and seeing patterns, a process begun in the early nineteenth century by **Humboldt**. Knowing the climate zone of a region can help decide what plants would be suitable to grow there, but also aid research about how particular adaptations have evolved. For example, plants that grow in the cool **deserts** of America, such as agaves, have much in common with the aloes, which grow in similar conditions in the Old World. But neither are closely related – there is convergent evolutionary pressure taking place, which shapes both groups of plants to adapt in similar ways. The main climatic zones on Earth are outlined in the Köppen climate classification, made by the German-Russian scientist Wladimir Köppen in 1884 and still the most widely used.

GARD CHRON.

Conservation

Conservation is a term which covers the many ways that we aim to protect and preserve the natural world. From the Latin *conservare* which means "to keep intact", it was a term first applied to the natural environment sometime in the nineteenth century, and was originally concerned with ensuring that the natural world was not depleted by humanity's actions upon it; a practical consideration, ensuring the continuance of resources. But in the latter half of the twentieth century, the term came to be much more closely associated with environmentalism, looking to the protection of the natural world for its own sake, and in its entirety.

The natural world is a complex web of interactions. The **extinction** of just one **species** can affect others in ways that we can't necessarily predict. As such, the priority is always what is called ***in-situ* conservation**, which aims to preserve the functioning natural world in place. This can be approached in several different ways: preventative actions such as legislation or **habitat** management, remedial actions such as habitat restoration.

But these interventions are imperfect and subject to the vagaries of the environment, natural disasters, ignorance of rules. So there is often a backup plan; that which is threatened is taken out of its natural habitat to be looked after elsewhere, hopefully with a view to eventual reintroduction. This is called *ex-situ* conservation and, for plants, might take the form of **seeds** stored in a **seed bank**, or living plants kept at a **botanic garden**.

Conservation relies on the belief that the natural world is worth protecting. This can be argued either from the view that the natural world is essential to humanity, or from the point of view that something in nature is valuable in and of itself for reasons that are not necessarily quantifiable.

Cyanobacteria

Cyanobacteria are a type of simple **photosynthesizing** organism. Being photosynthetic, they contain pigments that help them to absorb light energy including a form of chlorophyll, a pigment which appears green, and phycocyanin, which appears blue. This particular combination gives cyanobacteria both their scientific name (*kyanós* means "blue" in Greek) and their other common moniker, blue-green **algae**.

Cyanobacteria are microscopic, so they are visible only *en masse*, often as a colouring in water where they have rapidly multiplied into a bloom. But despite their small size, many play an important ecological role, converting atmospheric nitrogen into forms that other organisms can use. Certain cyanobacteria are being investigated for industrial applications as varied as the production of biofuels and **fertilizers**, and even in nutrition – spirulina is a type of cyanobacterium often consumed as a food supplement.

Desertification

Desertification describes the process of losing the plant life in an already dry area. Vegetation in dry places is often holding **nutrients** and water in place, and if it degrades, dusty **soils** progressively lose their ability to grow anything at all. Vegetation in dry **biomes** often acts as both a heat sink and a reservoir of water; plants affect the way precipitation happens, and control dust and sand aerosols that can feed into storm systems. Reducing vegetation increases the temperature of the land, further pressurizing the remaining vegetation. Desertification impoverishes human communities and reduces the biological richness of the land. Pressure from humans whose income is precarious is likely to create a feedback loop whereby more marginal land is brought into use, further reducing protective planting and increasing food insecurity. Such **habitats** go from being usable to vulnerable and, in worst case scenarios, unrecoverable.

Ecosystem

An ecosystem is a set of **species** of plants, animals, **fungi** and other micro-organisms, living as an interwoven entity in a particular habitat. In particular, the science of ecology describes the energetic relationships between different groups of living things: in the final analysis, who eats whom. Ecoystems are often depicted as pyramids with plants or **autotrophs** at the base, species that make their own food sources using solar (or less usually) chemical raw materials. Autotrophs are consumed by heterotrophs, organisms that consume primary producers. An ecosystem is a conceptual framework that helps us to understand how interdependent living things are, and how one thing becoming out of balance within such a system can threaten life for all organisms within it.

Each organism within an ecosystem is said to exist in an ecological niche, which means it has a position within that system for which it as a species is best adapted. This also describes how it relates to other species in the ecosystem, either by being in competition with them, or by predating on them, or by mutually cooperating with them in some way. Xerophytic plants, for example, are adapted to life in the niche of very dry habitats, where temperatures can fall low at night. **Cacti**, in the New World, and Euphorbias, in the Old World, have evolved similar responses to this niche, storing water in a modified stem, below a thick waxy **cuticle**, and having spines instead of **leaves** to avoid being eaten.

Endemic

This term is used by naturalists to describe **species** that occur in only one place. Because endemic species occur in a single specific locality, they are especially prone to **extinction**. The biological significance and richness of a particular location is sometimes suggested by reference to the number of its endemic species, that being the number becoming extinct if they were to die out in that locality. Areas with a high level of endemics have been identified as "**biodiversity** hotspots" – and must have at least 1,500 endemic species of **vascular** plants to qualify. There are currently 36 such priority hotspots, where global funding agencies concentrate **conservation** efforts.

The British Isles have relatively few endemics due to the ice ages, after which the land plants had to recolonize slowly. Isolated islands, however, are often rich with endemic species, as a few colonizers evolve into many new niches; about 90% of all species on Madagascar are known only there, so that ecologists sometimes refer to it as the "eighth continent". English endemic species are often known by the Latin name *anglicum*, and those endemic to Wales by *cambricum*; but an endemic species of hawkweed discovered in Wales in 2004 was named *Hieracium attenboroughianum*, after David Attenborough, by the taxonomist Tim Rich, who said: "This is a personal thank you for the years of fascination he has given me going to different places to search for new things."

Food security

Food security means a state in which all people have fair access to enough nutritious and palatable foods to live a normal active healthy life, and are not at risk of that access being removed suddenly. In recent years focus has come to rest on the threats to food security which might be presented by major disasters such as **climate** crisis, famine, drought or disease outbreaks. However, it is also important to consider how more regular challenges feed into food security, such as market and crop failures or the vulnerability of cash economies. Botanical research can help with breeding crops to be more robust in response to rapid shifts in temperature or water shortages during the growing season. International food policy needs to balance responses to both, to guarantee no one goes hungry.

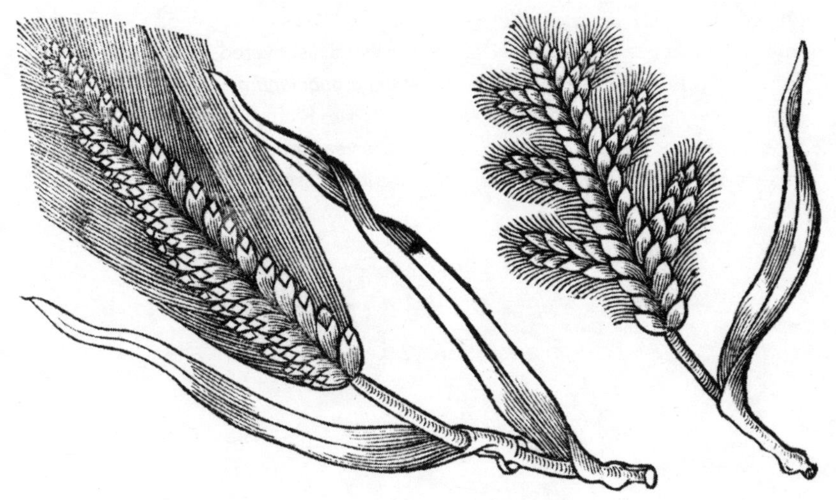

Fungi

Fungi comprise one of the kingdoms of life on Earth. It is only recently that we have understood how different they are to plants, and in so many ways closer to animals. While plants are **autotrophs**, able to feed themselves via **photosynthesis**, fungi and animals share a common reliance on feeding off other organisms to survive, making them heterotrophs. Unlike animals, though, fungi cannot move around to secure a food source, and have often adapted by evolving sophisticated symbiotic relationships with plants, to exchange biologically useful compounds. Today we increasingly recognize the role played by **mycorrhizal** fungi in aiding the survival of plants. Fungi are extremely significant for the continuity of life globally, both in their role as symbionts, and also in decomposing dead and dying things, releasing complex molecules back for use in the living world.

Fungi have been used in human culture throughout history, and predate it in the case of **hallucinogenic** "magic mushrooms". Their size does not indicate their significance; consider the penicillium mould from which Alexander Fleming made the first antibiotic. However, the largest living thing on Earth, recorded by *Guinness World Records,* is a single honey mushroom (*Armillaria ostoyae*) growing in Oregon's Malheur National Forest, covering 965 hectares, and estimated to be at least 2,000 years old and perhaps 8,000 years old. Nonetheless the fungi most useful to humans are most likely those in food production, ranging from bakers' yeast to the bloom on grapes that enables winemaking and the fungus used frequently in brewing, first named *Saccharomyces carlsbergensis*.

In-situ conservation

In-situ conservation is the attempt to preserve a **species** in its original **habitat**. This is a suitable way to proceed where the habitat is not threatened with complete destruction. It can also apply to efforts to conserve a whole **ecosystem**, which may in the long run be more successful, due to the complex interactions between individual organisms. National parks and **forests**, designated **floral** kingdoms and sacred groves are all examples of such care in preservation. Sometimes *in-situ* conservation can be combined with *ex-situ*, happening elsewhere: Kew ran a successful project to increase numbers of the St Helena ebony, an **endemic** species that had become extinct in the wild. Several thousand plants were cultivated by **micropropagation** at Kew and were then happily reintroduced to the island.

Invasive species

Invasive species are plants that have been introduced outside their natural ecosystems to places where they flourish, often at the expense of local natives. There is much debate about how to deal with invasive species, which are a part of an essential process in nature, as when new volcanic islands are colonized by species of plants and animals from elsewhere. Where invasives are deeply problematic, physical clearance can only be a stopgap. This leaves chemical or biological control as an option, both of which have their own serious impacts on the environment. Invasive species are an identified target of Kew's State of the World's Plants. The top five world invasives are:

Bromus tectorum, cheatgrass

Centaurea stoebe, knapweed

Phragmites australis, common reed

Phalaris arundinacea, reed canary grass

Pteridium aquilinum, bracken

Lichen

A lichen is an organism arising through the mutually beneficial relationship between an alga or **cyanobacterium** and **fungi**. The alga or cyanobacterium produces complex carbohydrates through **photosynthesis**, while the fungi absorbs minerals and water. But a lichen is the sum of more than its parts; both constituent organisms are changed by the **symbiosis** and become something else.

The root of the word is Greek for "to lick", but the Greeks primarily used it to mean a skin disease, which is probably the reason for this usage: lichen often appears as colourful, crusty growth on the **bark** of **trees**.

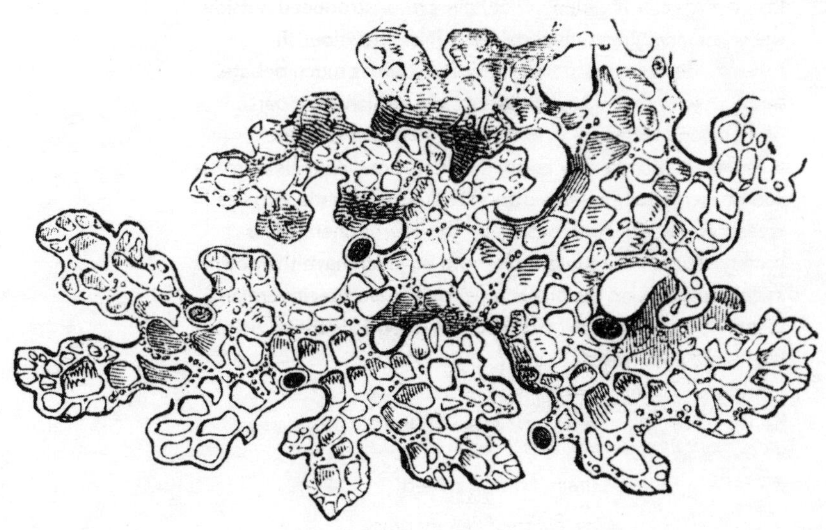

Mycorrhiza

From the Greek *myco* for "fungus" and *rhizo* for "**root**", mycorrhiza describes the coming together of plant roots and **fungi**. This important relationship is usually mutually beneficial; at a basic level the fungus aids the plant in its uptake of water and **nutrients**, in return benefitting from the more complex carbohydrates that the plant can produce through **photosynthesis**. There is some evidence to suggest that mycorrhizal relationships also play a part in plant-to-plant communication, and in resistance to disease, toxicity and drought. Some think that without the assistance of fungi, plants would never have been able to make the transition from water to land. The establishment of certain plants – most notably orchids, citrus and pines – is still entirely dependent on the presence of a mycorrhizal fungus.

Partner gardens

Kew works with a huge number of partners around the world, including many **botanic gardens** and national parks. **Conservation** projects have the best possibility of success when they are understood and positively led by local enthusiasts, preferably by botanists on the ground. However, such efforts must also seek funding in a competitive context, and conservation projects often focus on animals and much less on plants, so flagship botanic gardens such as Kew can act as a visible link between the two. For example, in 2015 Kew began to identify seven territories as Tropical Important Plant Areas. In the Caribbean, Kew is working with the National Parks Trust of the Virgin Islands to identify where grazing wild goats, farming and new roads are threatening British Virgin Island **endemics** such as the *Pitcairnia jareckii*, a bromeliad.

Soil

Though it comes from the Latin word *solum*, meaning "ground", soil is more than the earth beneath our feet. In fact, soil is defined as a complex mix of components: weathered rock and minerals, **organic** matter, gases and liquids held in pores. It is this combination which enables soil to support living organisms like plants, and thus makes it so vital for life on Earth. Ultimately, it is the foundation of almost all **terrestrial** food chains.

Soil also performs a host of other indispensable functions. It filters water sources, it can mitigate against flooding, it acts as a sink for excess carbon in the air. But the soil is fragile; modern agricultural practices can deplete soil nutrients, destroy its structure, and leave it vulnerable to erosion, ironically endangering the very thing on which we rely most for our food.

Weather

Weather is the atmospheric conditions of the natural world at any given time and place. If **climate** is what you might expect – the long-term seasonal patterns of temperature, precipitation, cloud cover, for example – weather is what actually happens. Affected by a whole host of variables, weather doesn't necessarily fit the season. Though we might expect February to be a cold, dreary month in the UK, we might well get some warm, sunny days which mean that we must think about watering plants in pots. And unseasonal weather like this is becoming more common as the climate changes. Along with extreme weather events such as droughts, heatwaves and storms, these unpredictable conditions can disrupt natural life cycles and **ecosystem** function and can impact food crops, causing problems in our food system just as much as the changing climate itself.

Wood wide web

A term to describe the way in which **trees** are understood in modern science to live as part of connected network of communication. Trees are increasingly seen in relationship to the organisms around them, including the **mycorrhizal** fungi that supply them with vital **nutrients** from below the **forest** floor. **Fungi** live in association with tree **roots**, and take from them essential carbohydrates in return for fungal breakdown products such as minerals. However, it is now understood that this relationship does not simply provide a tree with chemical compounds; it also links a tree into a complex system of communication about what is happening to other neighbouring trees. Trees are able to communicate directly using pheromones, and it is exciting to see how rapidly this field of understanding is developing.

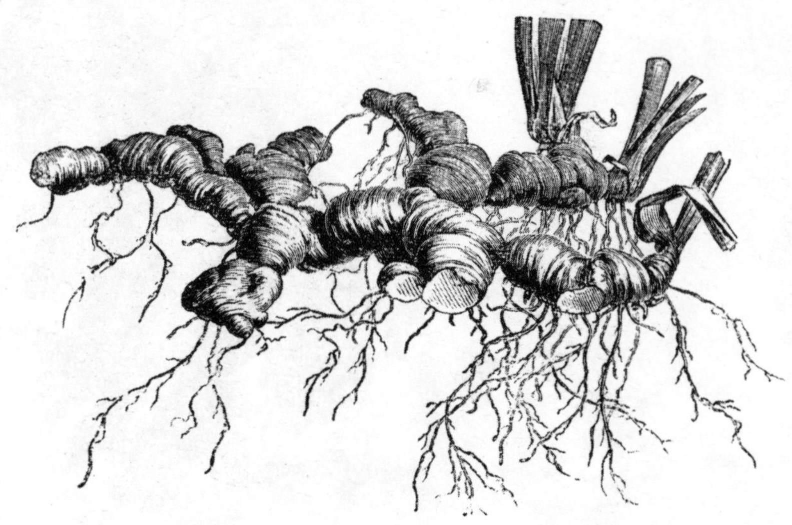

BIOMES & HABITATS

Antarctic

The Antarctic is a geographic region. It is the area surrounding the Southern pole of the Earth and is named for its geographic opposition to the **Arctic**, from the Greek prefix for "opposite", *anti*. Similarly to the Arctic, the Antarctic is a place of high winds, low rainfall and extreme temperatures. The lowest temperature ever recorded on Earth was in the Antarctic; -89.2°C (-128.5°F). The main landmass is the continent of Antarctica, of which only a tiny proportion is suitable for plant life. Even here, it is mainly simple organisms like **mosses** and **lichens** which predominate. There are, however, two flowering plants that have managed to carve out a place for themselves: Antarctic hair grass (*Deschampsia antarctica*) and Antarctic pearlwort (*Colobanthus quitensis*).

Arctic

The Arctic is also a geographic region. It is the area surrounding the Earth's Northern pole, hence the word's origin in the Greek word *arktos*, or "bear"; in Northern latitudes the constellation known as the Great Bear is particularly visible. The North Pole itself is in the middle of the Arctic Ocean, which is covered by a permanent ice sheet. The Arctic region is often defined as the area within a certain distance from this point. While the majority of Arctic landmass lies further south on the outer edge of this circle, conditions are still extremely hostile with low temperatures, strong winds and a permanent layer of frozen ground (the permafrost) below a certain depth, even in summer. Despite these extreme conditions, more than 1,700 **species** of plant still grow here, demonstrating the adaptability of the plant kingdom.

Biome

A biome is an area defined by the naturally occurring community of plants and animals which inhabit it. Major types of **terrestrial** biome include **desert**, **forest**, **grassland**, and **tundra** but definitions can be more specific, i.e., **tropical rainforest**, **temperate** grassland. Biomes are generally named for the plants that live in them because their animals are entirely dependent on the presence of plants. Around the world, similar environmental conditions produce similarities in adaptation. For instance, **grasses** tend to dominate in areas where rainfall is low, grazing animals are present, and there are regular wildfires. Both the American **prairie** and the Asian **steppe** have these kinds of conditions. While the particular **species** might be different in each of them, the kind of plants found in both (predominantly grasses) are similar, so both can be referred to as a grassland.

Bog

This is a wetland **habitat** where the **soil** is permanently saturated with water. This reduces the oxygen level, so that the decay of dead **organic** matter is slowed down, making it nutrient-poor and very acidic. Plants growing in bog habitats must be adapted to these conditions; typical **flora** include sphagnum moss, orchids, acid-loving shrubs such as heathers, and **carnivorous** plants, which have evolved to deal with the low nutrient levels by capturing and digesting animals and using them as a source of important **nutrients**. Boggy soils are unsuitable for farming but predominantly contain peat, a fuel and **growing medium**, which has made them vulnerable to commercial exploitation. The value of bogs as biologically rich habitats has only lately been realized and campaigners often struggle to get people to recognize their wild beauty.

Boreal

Boreal **forest**, or taiga, is one of Earth's major **biomes**. The word denotes the thick, tall, mainly coniferous **forest** occurring at high latitudes, where temperatures fall below zero for more than half of the year. This gives the **trees** a very brief growing season, and is such a harsh environment for growth that the range of tree species occurring is very narrow, although the **ecosystem** itself is very rich. Taiga is the largest land biome, and runs as a belt almost continuously around the globe across eight main countries, Canada, Norway, Sweden, Finland, Russian, China, Japan and the United States. Many of these boreal forests are unmanaged by humans, and remain fully wild areas. These forests are very significant in terms of the carbon reservoir, and store it on levels equivalent to tropical forests.

Canopy

Canopy describes the habitat in a forest that is above ground level. At this higher level, **tree** crowns provide a lightly shaded, moist growing environment for a rich variety of invertebrates, **mosses** and **lichens**. They also host larger organisms, such as epiphytes such as orchids, and plants that climb from a forest floor anchor, like lianas, which include the commercially significant rattan. The biomass represented by the canopy is extremely significant, and its productivity via **photosynthesis** is very high. Dominance in canopy terms means being able to grow taller than other surrounding trees, thus outcompeting them for light. Canopy trees incur significant costs to remain dominant, ranging from supporting the unwelcome weight of **climbers** to having to spend more resources on growing than on reproducing; it is not yet totally understood how they generally avoid touching their neighbouring tree (see **Crown shyness**, page 187).

Cape Flora

This term describes the distinctive **flora** that evolved at the southern tip of the African continent. Beyond the famous **fynbos**, the Cape Flora comprises several **habitats** and is the smallest of the six floral kingdoms of the world, which are areas recognized for their rare status globally, their very high level of **endemic species** and families, and their general richness of **biodiversity**. Some 12 plant families and 160 genera are unique to South Africa and only occur there; one of its distinctive local crops is Rooibos **tea**. The Cape Floral Region has been recognized by United Nations as a World Heritage Site, and balancing the pressure of tourism with the need to publicize and protect the area is a delicate and challenging task.

Desert

Desert is a kind of **biome**: an extremely dry area where only specially adapted plants and animals can survive. From the Latin *desertum,* meaning "abandoned", deserts are often thought of as barren places. They have a water deficit – more moisture evaporates from them than falls as rain – which produces very difficult conditions for life. But there are a host of organisms that have adapted to survive these extreme conditions; plants like **cacti**, for instance, have evolved fleshy stems to store water, and reduced **leaves** to limit water loss. A large proportion of desert plants have extremely short life cycles, which allows them to ride out unfavourable conditions as **seeds** and then quickly spring into growth and reproduce when rain does fall, producing brief but spectacular desert blooms.

Forest

Forest is a word with many different meanings, ranging from the amount of land that is covered, to the status of the **trees** in law, how tall the trees are and how far apart, and whether they have grown wild or are a plantation. In 2020 figures suggested that about a third of the world's land surface was forested, with more in the southern hemisphere than the global north, and more than half the forest on Earth is in just five countries - Russia, Brazil, Canada, China and the USA. **Tropical** forests are the most numerous kind of forest on Earth, representing about 45% of the global cover. As a whole, forest **habitats** contain more than three-quarters of the Earth's plants by biomass (weight). Despite their importance to human existence, about half of the world's forests are still essentially wild places, and forests occupy a special place in the imagination of human beings, from *Little Red Riding Hood* to *Game of Thrones* and *The Blair Witch Project*.

Fynbos

Fynbos, which means "fine plants" in Afrikaans, is found only on the southern tip of Africa. It is heathland or shrubland plant vegetation, on poor **soil**, and represents about 80% of the plant **species** of the whole **Cape Floral** Kingdom. There are about 9,000 species of plant growing in this very small region, an unusually high number in such a space, of which about 6,000 are **endemic** – they occur nowhere else. The heathland plants often have silvery, thickish, hard **leaves**, and eye-catching **flowers**, like *Protea*, but also include gloriously coloured heathers and *Leucospermum*. There are many **bulbs**, including an astonishing 96 different species of gladioli alone. The fynbos is a central global **conservation** priority: a single spot, Table Mountain, has 2,200 plant species, more than the whole of the United Kingdom.

Grassland

Grasslands are a type of **biome**. As their name suggests, they are areas where the plants are predominantly **grasses**. Grasslands are found where certain conditions mean that taller plants – particularly trees and shrubs – struggle to establish. Grasses, however, have evolved to survive these pressures, which include reduced (often seasonal) rainfall, grazing by animals and regular wildfires. Human actions can also play a role; cutting for hay or grazing by herds of domesticated animals can be the pressure that keeps trees and shrubs species in check, resulting in what is known as semi-natural grassland. Grasslands, both natural and semi-natural, support a huge diversity of plant and animal species but are endangered worldwide due to the spread of **agriculture** and the intensification of its practices.

Habitat

From the Latin *habitare* – "to inhabit" – habitat describes the place where an organism naturally lives. It is an often used and often misused term. Even within scientific literature it has been defined in numerous different ways. Generally, habitat is thought of as a type of environment – a chalk meadow or an oak **woodland**, for example. While this can be useful to give an idea of the kinds of places in which a species or community might be found, it is an over-simplification. Every organism has habitat requirements: certain basic conditions and resources which it needs from its environment. Its habitat will be any place where those requirements are met to the degree that it can survive and reproduce. For a plant, key factors might include light levels, water availability, soil type, temperature, even the presence of predators or pollinators. So when we say that the habitat of a particular plant is chalk meadow for instance, we are actually implying that a chalk meadow fulfils enough of those requirements to be that plant's habitat.

Island flora

Island flora is a collective term for the plant **species** found on islands. Islands are hotspots for **endemic** species – those not found anywhere else in the world. New Guinea and Madagascar (two of the largest **tropical** islands in the world) each have almost 10,000 plant species not found anywhere else on Earth. Geographically isolated by the ocean, their inhabitants evolve their own unique solutions to the problems of their environment and, because they do not cross over with any other species filling similar niches, diverge over generations to become utterly unique. But many island species are threatened. Not only because they only exist in very few geographical places, but also because islands by their very nature have limited space and resources, both of which are also required by an island's human inhabitants.

Mangal

Mangal is the word for mangrove **habitats**, where salt-tolerant **trees** grow along coasts, in water that is salty from mixing with the sea. The dense tree **roots** slow coastal wave action, and fine mud and sediment settles in the mangal, behind this barrier. This water source is a particular challenge to plants, which must filter the harsh salts out of water to avoid cell damage, but must also be strong and flexible in structural terms, to avoid wave damage. Mangroves have evolved special roots that grow upwards to avoid the challenges of immersion in salty water (see **Pneumatophore**, page 27). Mangroves are particularly important in terms of protecting softer sandy coastlines from erosion, and absorbing storm impact, lessening the impact not just on beaches themselves but also on other inland habitats. Much so-called blue carbon is stored underwater in mangrove habitats.

Mediterranean

The Mediterranean **biome** describes a maritime ecological niche, where summers are dry and hot, and winters see storms and much heavier precipitation on the whole, although temperatures are predominantly relatively mild. The biome is named after the Mediterranean, but it also covers the distinctive vegetation seen globally in this kind of habitat, for example in the Chilean Matorral and the Californian chaparral. Much of the characteristic vegetation has a waxy layer to protect against water loss, and hard woody **leaves**; typical species might be lavender or olives. The similarity of adaptation allows Mediterranean plants to be exported around the world: in modern times some of the most distinct vegetation of the Mediterranean are introductions, such as the American agaves that feature on many tourist posters celebrating southern France, and the **palm** trees of the Hollywood avenues, which have travelled in the opposite direction.

Miombo

The Miombo **biome** is a landscape of **tropical** and **subtropical** grass and scrublands in Africa, characterized by a degree of **tree** cover. The strongly identified trees are *Brachystegia* **species**, known in local languages as *Miombo*. Miombo forms a wide belt across the centre of southern Africa, growing typically on nutrient-poor **soils** and surviving for long months of the year on little rainfall. The trees are well-adapted to local conditions and host a rich variety of species. They do not lose **leaves** in the cool of winter, but instead shed them at the onset of the dry season, reshooting new leaf cover as the rainy season begins. They are heavily inhabited and used by humans, and have been identified as an important **conservation** target.

Montane

From the Latin for "mountain", the word montane indicates an association with mountainous areas. It is often used to describe kinds of biome which are found on mountain slopes – montane **forest**, for example. Temperature decreases with altitude, so the steep gradient of a mountain means that conditions on its upper slopes can be very different to the surrounding lowlands. As you ascend a mountain, the **climate** changes as if you were moving towards the poles. The kind of life you find therefore also changes, having adapted to these conditions. Montane **habitats** are essentially isolated from each other, surrounded as they are by lowland areas with a different climate. This prevents species from spreading, so many mountain plants are found only in specific locations.

Oases

Oases are contained fertile areas occurring in the **desert**. They form where water has been able to come to the surface from underground, via either natural or human-made means. They are an essential for life in the hostile desert environment. Many were first highlighted to human attention by vegetation, growing where migrating birds had stopped to drink, and humans have often worked hard to supplement the natural water supply by managing it carefully. Oases are often also carefully planted in layers to reduce the solar impact on crop plants growing below. Date palms generally provide the tallest element of cover with **fruit** and other productive orchard plants such as apricots and pomegranates growing in their shade.

Paramo

Paramo habitats are the world's most biodiverse high-altitude **ecosystem**, made up of moorlands in the tropical mountains of South America. They are an important conservation priority, because of their unusual balance of native **flora**, in particular distinctive giant rosette plants, such as puyas, and **grasses**. They have been seen as acting as a sort of sponge, taking in and storing water during heavy rains and then releasing it slowly to the benefit of local organisms. The paramo's high altitude means it often experiences extreme changes of temperature between day and nighttime, and the flora must be adapted to that, which has resulted in a high level of **endemic** species. Scientists have identified the paramo as areas where fast-changing **evolution** is happening, and **climate change** adds to the pressure on this rare habitat as its water patterns shift dramatically.

Prairie

From the French for "meadow", prairie is a word that has been adopted to mean the **temperate grasslands** of North America. The height of the Rocky Mountains creates an area of low rainfall on their Eastern side. Combined with the pressures of grazing animals and wildfires, these conditions mean that **grasses** and wildflowers are the dominant plants here, with characteristic **species** including Little Bluestem (*Schizachyrium scoparium*) and Black-eyed Susan (*Rudbeckia hirta*).

Much of the natural prairie has been lost. The fertile **soil** is ideal for **agriculture**, and this has been exploited to the extent that it is thought that less than 1% of the tallgrass prairie remains. A very similar type of **biome** exists in South America, where it called *pampas*; in Eurasia, where it is called *steppe*; and in Southern Africa, where it is called *veldt*.

Rainforest

Rainforest is a type of biome: an area dominated by **trees**, where a hugely diverse plant and animal community is shaped by high levels of rainfall. While the term is perhaps inexorably linked with **tropical** rainforests like that of the Amazon basin, there are also temperate rainforests such as the coastal **forest** of California, and the Caledonian forest in Scotland.

Both tropical and temperate rainforests are perhaps the most visible of endangered **ecosystems**. Whether through logging, clearance for agriculture or development, or multiple other causes of deforestation, this vital **habitat** continues to disappear at an alarming rate. And this is not just a loss of trees. Rainforests are home to such a diversity of life that huge numbers of **herbaceous** plants, animals and insects are also threatened by their loss.

Riparian

Riparian **habitats** border running water such as rivers and streams. **Aquatic** plants are very adapted to these conditions, which present several special challenges. The first of these involves access to light and gas exchange. Where one side of the **leaf** or both is in water, this makes the normal use of **stomata** impossible. Additionally, watery living presents a challenge for the distribution and **germination** of **seed** because plants must find a way of getting their seeds to somewhere with access to light, water and air. Lastly there are the structural problems of growing in moving water, which require particular adaptations to conventional **roots** and stems, often to hold air so that the plant stays buoyant in the water. Additionally, leaves are often floating, like the famous Amazon waterlily, *Victoria amazonica*.

Saltmarsh

A saltmarsh is a community of plants which exists on land regularly flooded with salty water. Saltmarshes are found in temperate latitudes, in flat tidal areas where sediment has built up to the extent that plants are able to colonize it, often around the mouths of rivers. Any plant that grows in these places must be specially adapted to cope with the salty conditions and is called a **halophyte**.

Sometimes considered uninspiring to look at, saltmarshes are in fact extremely important and productive ecosystems. Studies have shown that they photosynthesize as much energy as an equivalent-sized area of tropical rainforest. They are permanent home to a host of unique species, and provide a temporary nursery habitat for many more, including some commercially important fish. But more than this, saltmarshes stabilize and protect coastlines, absorbing flooding and storm energy, acting as a buffer between the sea and the land.

Savanna

Savanna is a type of **biome**. Although it stems from a Taino word meaning "treeless plain", a savanna is actually a landscape where **trees** and **grasses** co-exist. Usually, the presence of trees makes it difficult for grasses to grow; the **canopy** acts as a shade, blocking out sunlight vital for healthy grass growth. But in a savanna, the trees are both widely spaced and of a type that lets direct sunlight reach the ground underneath, allowing grasses to grow as well. This balance is maintained by extremely seasonal rainfall and periodic wildfires, along with the presence of migratory grazing animals. Savanna can be found on four of the world's continents, but perhaps the most well-known are in Africa, where the unique conditions have produced distinctive trees such as the baobab (*Adansonia* **species**), and *Acacia*.

Subtropical

This term is used of plants adapted to living in the subtropical zone, immediately north and south of the tropics of Cancer and Capricorn. Usually frost-free, the subtropics include very diverse vegetation, from **savanna** and true **desert** to hardwood **forest**. Despite their high latitude, these conditions can be demanding because temperatures often fall at night. Subtropical vegetation can also occur at **tropical** latitudes, where they are high in the mountains. **Palms** and other lush **fruits**, such as mango and avocado, flourish in the gentler parts of the subtropical zone, requiring long hours of sunshine and sufficient rain. Subtropical vegetation has a significant history in horticulture, often used to evoke an "exotic" feeling by combining in one flowerbed plants from all over the world.

Swamp

Swamp **habitats** are ones permanently soaked with water; they can be freshwater, saltwater or brackish (a mixture). Marshes and swamps in botanical terms are different: marshes contain running water sources, whereas swamp forms around still water, supplied by tides, flooding or groundwater. Swamp vegetation is specially adapted to absorb essential elements such as nitrogen and phosphorus. Swamps have historically been perceived to be both unproductive and also likely to cause disease via insects, but in recent years their value as habitats, and also in flood and coastal management, has been much better recognized, and swamps are being restored where possible. Swamps are often beautiful and atmospheric places but lack the aesthetic draw of other forests, and even the biggest ones are not well known: the largest in America is evocatively titled the Great Dismal, covering more than 1,940 square kilometres, and the little-discussed Vasyugan Swamp in Siberia is larger than Switzerland.

Temperate

In botanical terms, temperate describes the regions above the tropics but below the poles, where more seasonal **weather** occurs on a changing basis through the year. The climatic patterns, and thus the vegetation, are determined by a range of factors including latitude, but also rainfall, wind, altitude and position on the continent. The zones that count within the category of temperate are **subtropical**, which comprises both humid and monsoon **climate** systems, and have long wet summers; **Mediterranean**, which have dry summers and heavier winter rainfall; **subtropical** highlands, which tend to be mild year round; and oceanic climates such as New Zealand and the United Kingdom. There are also a set of temperate subpolar climates which experience four seasons with a very strong winter: Alaska and Chile both belong in this zone.

Terrestrial

Terrestrial means plants that dwell on land, as opposed to in the water. Lots of plants can tolerate being flooded for short periods of time, and there are **prairie** and meadow plants that grow best with a period of inundation at least once a year. However, most terrestrial vegetation represents a distinct development in the **evolution** of green plants, which happened around half a billion years ago. At this point, some green **algae** began to adapt to live out of the water, evolving ways of dealing with the greater variations in temperature on land and the higher light levels. From here, liverworts and **mosses** split from the other plants, and later **ferns** and horsetails, while the last to evolve were the flowering plants, about fifty million years ago. The significant later adaptations for life on land include the structural elements of **vascular** tissue and true **roots**, so that transport, anchoring and growing upwards were possible.

Tropical

Tropical **climatic zones** experience warm growing conditions all year round, with strong sunlight and often abundant rainfall, and are usually defined as having monthly average temperatures above 18 °C. Plants must manage these growing conditions, which would appear ideal and yet present substantial challenges, in terms of the intense solar radiation alone. Tropical vegetation tends to be divided into three zones: **rainforest**, monsoon and **savanna**. With many economically important crops grown here such as **banana**, coconut, oil palm, bamboo and teak, tropical **habitats** represent a huge productive zone, but are also a globally significant carbon sink and **biodiversity** hotspot. However, because they are such a rich growing environment for lucrative crops, there are serious threats to the tropical **climate** zone and its vegetation; rainforests in particular house about half all the living **species** on Earth.

Tundra

Tundra is a type of **biome** found in cold, dry areas of the world. From the Finnish word *tunturi*, which means "a treeless plain", tundra begins at the point where conditions become too extreme for **trees** to survive. The plants that make up tundra vegetation are specially adapted to these conditions. They are low growing to resist the wind and to be closer to the ground where they can take advantage of the insulation offered by snow, they have narrow **leaves** to reduce water loss. The summer season when temperatures rise above freezing is extremely short, and even then, the ground only thaws to a certain depth while the lower layers stay permanently frozen: the permafrost. Tundra are found near to the poles, where they are called **Arctic** or **Antarctic** tundra, or at high altitudes where they are called **alpine** tundra.

Woodland

A woodland is an area of land dominated by trees. It is an old word, an adaptation of the Old English *wudulond*. But it is a word whose precise definition is difficult to pin down, often used interchangeably with similar terms like **forest** or wood. However they are named, areas dominated by native trees and are particularly vital for biodiversity in the UK. Many UK native species arrived here with the end of the last ice age, including tree species that quite quickly colonized most of the landmass. A great many of our native plants and animals therefore evolved in a woodland environment, becoming dependent on the trees that made it up. For example, the UK's two native oak species alone support up to 229 species rarely found on any other tree. The loss of woodlands – now reduced to just 13.5% of the UK landmass – therefore risks more than just a loss of trees.

Science

Alkaloid

Alkaloids are a particular type of natural compound. Originally named for their perceived similarity to alkalis (the chemical opposite of an acid), they are produced by some select **fungi**, animals and insects but are most associated with the plant kingdom. It seems likely that they evolved as a defence against predation. Alkaloids have a bitter taste, and many are poisonous or at least have a significant physiological effect on any animal consuming them. Many alkaloids have been harnessed by humans for this very reason. They are used in medicine; **quinine** is used as an antimalarial, ephedrine to treat asthma, and morphine as pain relief. Many others have been appropriated as recreational drugs; caffeine, cocaine, and nicotine are also all alkaloids.

Anatomy

Plant anatomy is the study of the internal structure of the plant, while **plant morphology** is the field that focuses on the external structure and form of the plant. Much plant anatomy is done at the microscopic level. Plant cells have a set of typical features, beginning with a strong cell wall, which is outside the membrane that edges the cell itself, and is comprised of robust molecules, especially cellulose. Inside, the central area of the cell has the ability to store starches made via **photosynthesis**, for energy. It also often has a reservoir called the vacuole, from which water can be moved to maintain rigid pressure inside the cell, and which is also used for storage of waste and useful products. The energy block of the plant cell is the chloroplast, containing the necessary equipment for photosynthesis.

Astrobotany

The study of plants in space, astrobotany is a field designed to investigate how plants might be best enabled to flourish during space travel and on other planets, in order to facilitate longer-term voyages and stays in space. Plants are essential to humans both as food and as converters of carbon dioxide back into oxygen, regulating the atmospheric balance. Plants can suffer in a space environment, as a result of receiving non-ideal levels of solar energy and also because of the heightened levels of other kinds of radiation, from which our **atmosphere** normally somewhat protects us. The first plants to be grown to flower and produce **seeds** in space were from *Arabidopsis thaliana*, or thale cress, on the Salyut 7 space station in 1982, but Expedition 40 to Soyuz managed to grow red Romaine lettuce. The cosmonauts were not, however, permitted to eat them, and the lettuces were frozen and returned to earth for testing.

Autotroph

A word borrowed from the German and based on ancient Greek words meaning "self-nourishing", autotrophs are organisms that produce their own energy. This is in contrast to heterotrophs, which must consume other organisms to sustain themselves. Most autotrophs – including plants, **algae** and **cyanobacteria** – convert light energy from the sun into chemical energy through a process called **photosynthesis**, but organisms like deep sea bacteria which rely on chemical processes to produce their energy are also considered autotrophic. All animals including humans are heterotrophs, which means that we rely entirely on autotrophs for survival – they are at the base of every food chain and are therefore also called primary producers. A by-product of photosynthesis is oxygen, further raising the importance of the plants and their cousins to all life on Earth.

Carbon cycle

The carbon cycle is the natural process by which the element carbon moves through Earth's spheres. Carbon is the basic element of life, all known organisms contain it and it is an essential building block of many vital **organic** compounds. So, a major part of the carbon cycle is its movement through the biosphere. Plants play a crucial role: carbon in the atmosphere only enters the food chain when taken in as CO_2 by plants and other photosynthesizing organisms.

Under the right conditions, the carbon in living organisms is temporarily removed from the cycle when they die and converted by geological processes to become fossil fuels. When these are burnt, they release carbon back into the atmosphere as CO_2. The rate at which this is happening is causing an imbalance in the carbon cycle which is at the root of two of our most pressing environmental problems: ocean acidification, and **climate change**.

Crown shyness

This is the term for how even in dense **forest** of certain **species**, neighbouring **trees** are reluctant to touch one another, and will stop growing just short of doing so, leaving visible channels between their **leaves** and those of a neighbour. Not all trees do this, but it has been noted in pines, larches and some eucalypts. There is some research evidence that this is a behaviour which has evolved to stop predatory pest insects easily spreading, although this has not been proven. Another theory is that plants can sense when their leaves are getting close to the leaves of other trees, by using photoreceptors adapted for this purpose. But some argue it is simply caused by trees rubbing up against one another when blown by the wind. However it happens, a forest exhibiting crown shyness is, when viewed from above, a magical sight.

Dendrochronology

Dendrochronology is the science of analyzing growth rings in trees. It is also called tree-ring dating. As well as growing taller and producing new shoots, woody plants expand outwards each year, growing thicker. In cross section this growth is visible as concentric rings, each of which corresponds to a year of growth. By counting these rings, it is possible to work out how old a tree is. Analyzing the pattern of growth can also tell us about conditions during that tree's lifetime; a dry summer or a cold winter will result in a thinner ring, a wet summer or a mild winter in a thicker ring. Not only does this provide historical **climate** data but comparing patterns of growth in different trees also makes it possible to date human-made wooden structures.

DNA sequencing

Until the advent of cheap rapid DNA sequencing, the relationships between **species** of plant could only be seen visually. Sophisticated **taxonomy** developed, but with the advent of DNA studies botanists can show the relationships between species at a genetic level. When first invented, DNA sequencing was laboriously slow, but has been a serious possibility for plants since the invention of the PCR or polymerase chain reaction, allowing a degree of automation in processing. This technical innovation has enabled a global group of botanists called the Angiosperm Phylogeny Group to make for the first time a family tree of flowering plants, showing exactly how closely related the members are. Many herbaria have already changed their organizational systems to reflect the new family tree; Kew, for example, has entirely redesigned its plant family beds into an "evolution garden", to reflect the new scientific insights.

Ethnobotany

Ethnobotany derives from the Greek word for "nation", *ethnos*, together with botany. It is the field that studies the way in which people around the world have used plants, in practices ranging from **medicinal** through to religious, and sometimes both at once. The medicinal uses of plants fall into a particular specialist field called ethnopharmacology, which has been a source of rich indigenous knowledge to modern Western medicine at times. For this reason, intellectual property rights are now a central issue in the field and most countries globally are signatories to the Convention on Biological Diversity. This treaty specifies particular behaviours around the equitable sharing of the profits and benefits from the use of biological resources. Beyond medicine, though, ethnobotany might show how native peoples used **resin** to waterproof their boats, or provide useful knowledge about food **crop wild relatives**.

Evolution

Evolution is the process by which new species evolve, but also by which existing species adapt to changing environments, and indeed how whole new branches of the plant family tree can evolve in response to new conditions. The study of plant evolution is of primary significance to botanists, allowing insight into how the structures and functions of plants developed, and why. Knowing a plant's family tree can potentially help researchers looking at the medical potential of a toxic species; scientists can examine whether there are near relatives, possessing similar biologically active substances but less poisonous.

In recent years one area of significant focus in the study of plant evolution has been how **flowers** and insects evolved together. This is co-evolution, as animal **pollination** became the most significant way of fertilizing flowering plants. Plants evolved **nectar** to ensure pollinators would visit, and sometimes also evolved structures that favour particular pollinators. An example of this is the hammer orchid, a highly deceptive species that has evolved to evoke the female thynnid wasp. They are pollinated by male thynnid wasps, tricked into believing the flower is a possible mate. This behaviour – orchids mimicking female insects to trick pollinators into "mating" with them – was first discovered by Edith Coleman in the 1920s, but it is now understood that the resemblance is not visual but is achieved by mimicking the wasp's pheromones.

Insects, on the other hand, evolved ways of cutting their own costs, preferring to avoid any work the plant needs them to do which is not essential to them. Brassicas have repeatedly evolved different off-putting oils to stop butterflies laying eggs on their **leaves**, but caterpillars have in response evolved ways of digesting and breaking the oils down, in an "evolutionary arms race".

Extinction

Extinction is when a functional reproductive population called a taxon or **species** can no longer reproduce itself and dies out. There are currently about 9 million species of complex organisms alive on Earth, but over the 4 billion years of life, there have been more like 5 billion species alive at some time or other. Over the course of global history, many millions of species have become extinct, as **habitats** shift and adaptations become less advantageous. These small extinctions might pass almost unnoticed in the fossil record. Sometimes in Earth's history large-scale mass extinction events have taken place due to large-scale and rapid environmental change, especially as measured in geological time. In the twenty-first century we face what some ecologists have called the Sixth Extinction, the anthropogenic Holocene or human-made extinction.

GM crops

Genetically modified crops are plants whose DNA has been altered by humans using specific methods. These involve utilizing biotechnology, rather than plant breeding, to introduce or remove specific genetic elements. The first genetically modified food to go on sale was a 1994 tomato called the Flavr Savr, which was intended to increase the fruit's shelf life. GM crops have been the subject of enormous controversy because of a sense of concern about transposing elements between **species**. However, Kew scientists have focused on a few areas where important breakthroughs might be possible using gene editing. Gene editing only rearranges genes from within the organism's own DNA, and introduces nothing from outside. For example, clonal crops such as **bananas** can have areas of vulnerability to fast-spreading diseases "edited out".

Hormone

Plant hormones are substances produced within plants and which control many aspects of a plant's biology. From the Greek verb *ormán,* which means "set in motion", hormones precipitate chemical changes at a cellular level which, in turn, can affect growth, reproduction, even **senescence**. Many hormones are produced in one place, but travel within the plant to have an effect elsewhere, so they are often seen as messengers: the plant's internal post system.

Plants are particularly dependent on their hormones to respond to their environment. An animal can run away or fight back if attacked, but a plant's only option is in its hormones. For example, in many plants, a wound will result in an increased production of hormones called jasmonates. In different species these can variously signal cells to produce toxins, form insect-resistant barriers, or even to become less digestible – all responses to deter a herbivore from attacking again.

The power that hormones have over plants has been appropriated by humans. Many gardeners use hormones to encourage **cuttings** into producing **roots**, for instance. But another plant hormone is even more ubiquitous. Ethylene plays an important role in the ripening process of certain **fruits** including apples, bananas, and mangoes. Knowing this, commercial growers often choose to pick their fruit early, making it easier to transport and keep fresh. The unripe fruit is stored in an environment that limits ethylene production, and so delays ripening. When the fruit arrives at its destination it is allowed to start producing ethylene once more, or a synthetic substitute is applied. Thus, it arrives on the supermarket shelf perfectly ripe.

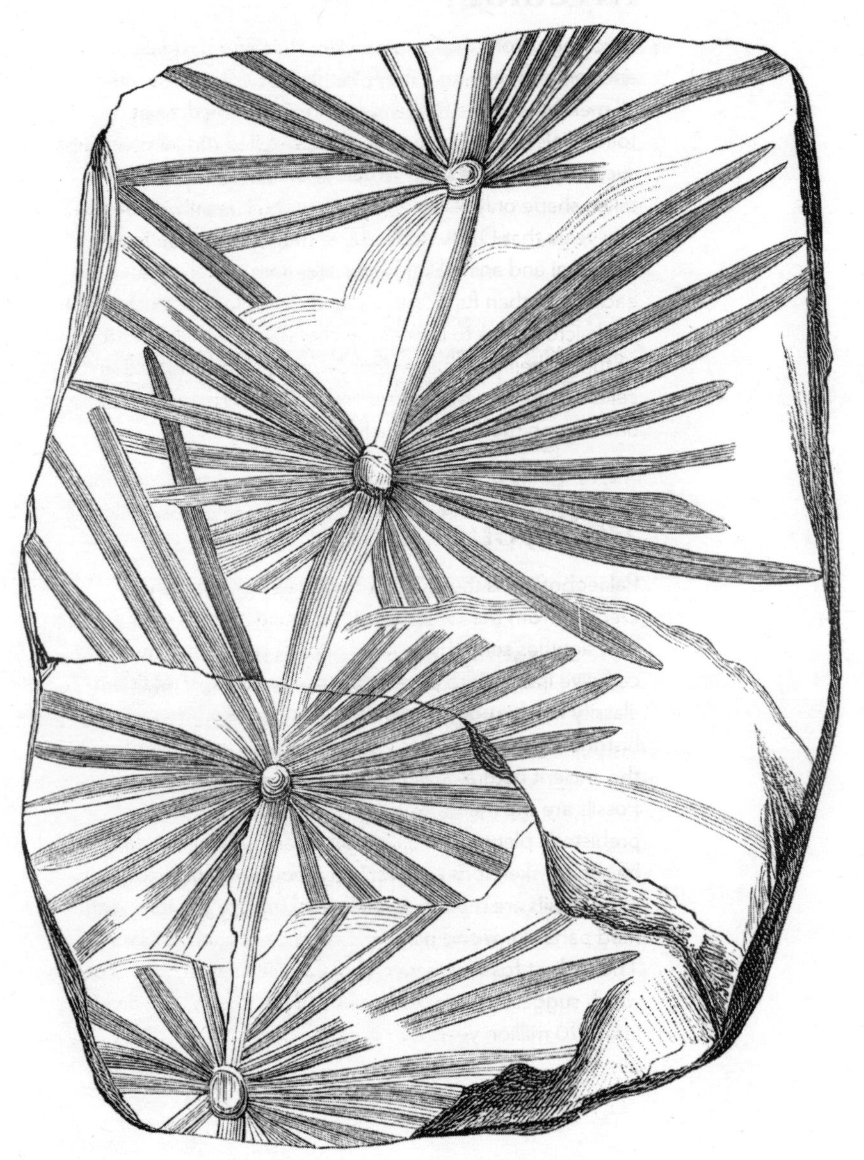

Mycology

Mycology, from the Greek *mukes*, meaning "fungus", and *-ology*, meaning "study", is the field concerned with all members of the fungal kingdom, from the dramatic toadstools of fairy tales to the single-celled fungal organisms such as yeast. The study of **fungi** as a unified field has taken shape only relatively recently, and it is only in the last few years that DNA studies have shown that genetically the fungi and animal kingdoms are more closely related to each other than fungi are to plants. Mycological study might use microscopes to look at the characteristic cell structures of fungi: their cell walls are made up of chitin, rather than cellulose, to give them structural stability; it is a relatively uncommon molecule in the living world.

Palaeobotany

Palaeobotany is the study of fossilized plant remains. Derived from the Greek word for "ancient", *palaeo-* is a prefix that signifies something of the ancient past. Ancient plants can give insight into plant **evolution**, helping scientists to classify modern-day plants. But they can also give a sense of historical **climate** and ecology, helping us understand what the ancient natural world was like and how it functioned. Fossils are the means by which scientists study these prehistoric plants. But, unlike lots of animals, plants do not have hard skeletons that last long enough to fossilize. So plant fossils are more rare and less complete a picture, with hard parts like **wood** being the most commonly preserved. The earliest fossil evidence of plants are microscopic **spores** which suggest that **terrestrial** plant life began to evolve at least 470 million years ago.

Phenology

Phenology is the study of timing in the natural world. From the word "phenomenon" – in the sense of something that happens and is observed – phenology examines the occurrence of events or stages in an organism's life cycle which happen on a regular basis, particularly looking at how their timing is affected by **climate** and **weather**. In plants, this might include when their **flowers** open, when their **fruit** forms or when they drop their **leaves** in autumn. Understanding what affects the timing of these events can allow us to manage them, which is particularly useful for growers. But phenology has wider applications; these kinds of events are extremely sensitive to – indeed, reliant upon – environmental conditions. Phenology can therefore provide scientists with concrete data on how the natural world is responding to the world's changing climate.

Photoperiodism

Photoperiodism is a response to changes in the relative lengths of day and night. It can be observed in many organisms, but it is particularly prevalent in plants. Some respond to the longer periods of darkness in winter and are called short-day plants. Some respond to the shorter periods of darkness and longer days of summer and are called long-day plants. A particularly vivid example of photoperiodism can be observed in *Poinsettia* **species**. These plants are so associated with Christmas because they will naturally produce their most striking feature in the middle of winter; *Poinsettia* are short-day plants and must be without light for at least 14 hours in each day for several weeks before they will produce their flowers and the bright red bracts that accompany them.

Photosynthesis

Photosynthesis, from the Greek *phos*, for "light", and *synthesis*, for "combination", describes the primary productive process of life on Earth. Plants, **algae** and photosynthetic bacteria all use specially adapted areas in their cells to absorb light energy from the sun, and convert it into storable usable energy, in the form of sugars and starches. When photosynthesis began on Earth, it greatly increased the atmospheric level of oxygen. This then made it possible for other life forms to evolve. These secondary producers did not photosynthesize but instead consumed primary producers, and breathed in oxygen to let them make use of their energy.

The parts of the plant cell which participate in photosynthesis are called the chloroplasts, which have inside them "reaction centres", containing the green pigment chlorophyll. However, the energy transfer from light is only the first part of a complicated series of reactions called the Calvin cycle, which bind in more carbon dioxide to produce stable carbohydrates that the plant can store. Solar energy is damaging as well as useful to plants, and plants must have photo protection to stop their exposed cells being damaged, just as we wear protection from bright sunlight.

The average global rate of energy captured by photosynthesis is 130 terawatts; this is currently eight times what human beings require. The use of carbon this way does not just enable us to breathe and eat; it also keeps the atmospheric level of carbon dioxide low, preventing **climate** warming. Increasing the amount of plants on Earth is one important tool in the armoury to deal with climate crisis.

Phylogeny

Phylogeny is the study of the family relationships between plants. It creates a branching tree which shows the **evolution** of plant families over time, splitting progressively from a single common ancestor. In the past this work was done by looking at the physical structures of plants, but today **DNA sequencing** is a powerful and invaluable tool in helping reconstruct these relationships with much greater accuracy. Work done by Kew in partnership with other **botanic gardens** worldwide, as the Angiosperm Phylogeny Group, has elucidated the science greatly. The process has led to some surprising removals and changes to the existing family tree of plants, which can be frustrating to gardeners when familiar names change.

Phytochemistry

Phytochemistry, from the Greek *phyto* "relating to plants", is the field dealing with chemicals made by plants. Plants produce phytochemicals to feed themselves, defend themselves and help themselves reproduce, and simply as by-products of other reactions. Some of the most useful to humans are those generated by plants for their own protection, as they tend to be strongly pharmacologically active, such as **alkaloids**. This group includes substances such as nicotine, caffeine, cocaine, opium and **quinine**, as well as less infamous substances such as important chemotherapy drugs. Many plant chemicals are defensive, and represent innovations in a struggle between producer and grazer referred to as an "evolutionary arms race", as each side adapts to try to outwit the other. The terpenes that give pine trees their distinctive turpentine smell effectively protect the **trees** against insect damage.

Phytogeography

Phytogeography is the study of where plants grow in the wild. By collating data on where a **species** or community has been observed or collected, scientists can build a picture of its pattern of distribution. This is useful in itself – for defining **conservation** areas, for example – but the key aim of phytogeography is to understand the reasons behind this distribution. Where a particular plant is found or not found can reflect not only the requirements and adaptability of that species, but also its evolutionary history. When combined with where similar plants have previously been found (for phytogeography can include historical data including fossil evidence) plant distribution can even help scientists infer such things as past movement of continents and tectonic plates, or historical **climate change**. So information on a plant's physical place in the world can be a very informative thing indeed.

Plant blindness

This is a particular challenge faced by conservation efforts, because research has shown that humans are more skilful at identifying and responding to images of animals than of plants. The term was coined in 1998 by Elisabeth Schüssler and James Wandersee, American botanists, to typify the difficulties faced by educators in simply getting human beings to see plants in the local environment. As Schüssler has said, "humans can only recognize (visually) what they already know", and it has been acknowledged that botanical skills are difficult to maintain. As a species we tend to be most interested in things most like ourselves; this skewing is demonstrated by emojis, where there are more than a hundred recognizably different animal emojis, and less than ten for specific kinds of plants. Effective apps for plant identification definitely present an opportunity for change and growth in this area; many users say that the apps are not tools for passive use but rather are tools that enable permanent learning.

Plant morphology

This is the study of the external form a plant takes, ranging from the way it unfolds its seed leaves from the germinating **seed**, through to the distinctive shape its reproductive parts take. Pollinators for flowering plants may range from tiny insects such as ants, to larger animals like hummingbirds and bats, and particularly refined nectaries have evolved in **flowers** such as the Madagascar star orchid, which is fertilized by a single species of moth. When **Charles Darwin** first saw the orchid, he predicted from the length of the **nectar** tube that the moth pollinator would have a tongue 30cm long: "Good heavens, what insect can suck it?!" he wrote. The hawkmoth itself was identified in 1903.

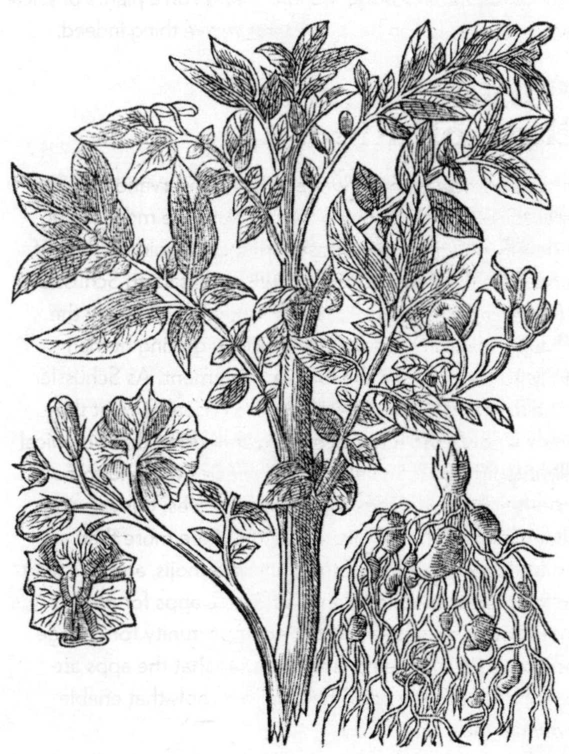

Plant pathology

Plant pathology is a significant area of research for gardeners and food-growers who are challenged by diseases such as rusts, and blights. The science concerns itself with infectious agents and environmental conditions, but not insects and other **pests** that predate the plant. The majority of plant pathogens are **fungi**, but plants are also vulnerable to bacteria and viruses, as well as to a less well-known group of biological organisms including virus-like organisms, protozoa and nematodes. One of these is the oomycetes, the infectious agent behind the failure of potato crops in Ireland which marked the start of the Great Famine. These still cause significant loss as tomato and potato blight. Most such plant diseases begin by breaking down the tough plant cell wall in order to penetrate the rich resources within.

Plant physiology

The investigation of how plants begin, grow and reproduce, plant physiology studies how plants grow initially from **seeds** or are propagated, and how they can best be nourished. It looks at how they get access to gases and water and light, how they use these to photosynthesize, and how they store the sugars and starch produced. It studies the reverse part of that process, when plants take in oxygen and use the stored carbohydrates as energy for cell reactions. It investigates how plants move their **nutrients** around. And it looks at how they reproduce, including all the systems involved. Plant physiologists can use their findings to make intelligent suggestions to commercial growers – for example, how to successfully **force** orchids into flower for supermarkets.

Plant pigment

There are three principal groups of plant pigments, comprising most of the coloured hues in the plant kingdom: chlorophyll, carotenoids, and the flavonoids, which comprise the anthocyanins and the anthoxantins. The most central pigments to plant life are the chlorophylls used for **photosynthesis**. Chlorophyll heavily absorbs blue and red light, explaining why almost all plants appear green to our eyes. Carotenoids are supplementary pigments for photosynthesis, to make sure as much light energy is harvested as possible, but they also help protect the plant against the harmful effects of solar radiation, called photooxidative damage. Xanthophyll is a carotenoid, and creates the dark-green colour seen in kale and broccoli. The carotenoids are the most important as colours in economic terms, and are often used to create vivid tints in food.

The final group, the flavonoids, are important in attracting pollinators, providing colours that are attractive to animals. The flavonoid anthocyanins are fluorescent under UV light, producing a range of additional colour impressions visible to pollinators able to see that part of the light spectrum. Anthocyanins, from the Greek *anthos* for "flower", and *kyaneos* for "dark blue", are pigments of many shades that create most of the familar bright red, purple and blue colours of **fruits**, **leaves** and **flowers**. Bright colours are useful to attract pollinators but anthocyanins may also protect against certain kinds of stress. They are particularly high in berries such as myrtles, redcurrants and blueberries. They also create the bright colours associated with the best autumn leaf colour.

Tannins

Tannins are a type of chemical compound produced by many plants. They act primarily as a deterrent to herbivores, reacting with proteins in saliva to produce a dry and bitter feel in the mouth, which puts animals off from feeding on the plant. A more pleasant equivalent of this sensation can be experienced with some red wines – tannins are present in grapes and can also be absorbed from the oak barrels in which wine ages, giving some wines a dry, oaky texture. The word *tannin* relates to the role of these chemicals in turning hide into leather; a process known as tanning, and which traditionally uses the tannins present in tree **bark**. Tannins are also used in several other industries, including ink manufacture, dyeing and oil drilling, and can be found in some medicines.

Taxonomy

Taxonomy is the branch of science which classifies the living world. Stemming from a combination of the Ancient Greek words for "order" (*taxis*) and "law" (*nomos*), taxonomy defines how similar organisms can be grouped together. As such, it is the scientific thinking behind groupings of plants like the **species** *Bellis perennis,* or the mint **family**, Lamiaceae. Even the plant kingdom itself is a taxonomic grouping.

For our early ancestors, being able to recognize similar and different kinds of plants meant the difference between sustenance and death. But taxonomy has become essential in our attempts to understand the world, not just to survive. It is how we record the diversity of life, the framework within which we discuss the natural world.

In taxonomy, any grouping of organisms is called a taxon. Taxa (the plural term) are given a rank according to how similar their members are based on shared characters. The lowest rank is species and they ascend through **genus**, family, order, division, phylum, kingdom, to domain. Taxa at each rank are grouped together to form a taxon at the rank above, resulting in a hierarchy of classification. Any organism will be in a group at every level, giving us multiple different ranks in which to consider it.

For a long time, classification was based solely on appearance; plants which looked like each other were grouped together. But taxonomy seeks to reflect the evolutionary relatedness of organisms, and this aim has been helped by the vast expansion of information available, most notably through DNA analysis.

Index

Credits

All images are taken from the Library and Archives collection of the Royal Botanic Gardens, Kew.

Every effort has been made to acknowledge correctly and contact the source and/or copyright holder of each picture. Any unintentional errors or omissions will be corrected in future editions of this book.